THE EXPLC
OF SPACE

ARTHUR C. CLARKE

B.Sc., F.R.A.S.
Chairman, British Interplanetary Society

HARPER & BROTHERS   PUBLISHERS   NEW YORK

Library of Congress catalogue card number: 52-5430

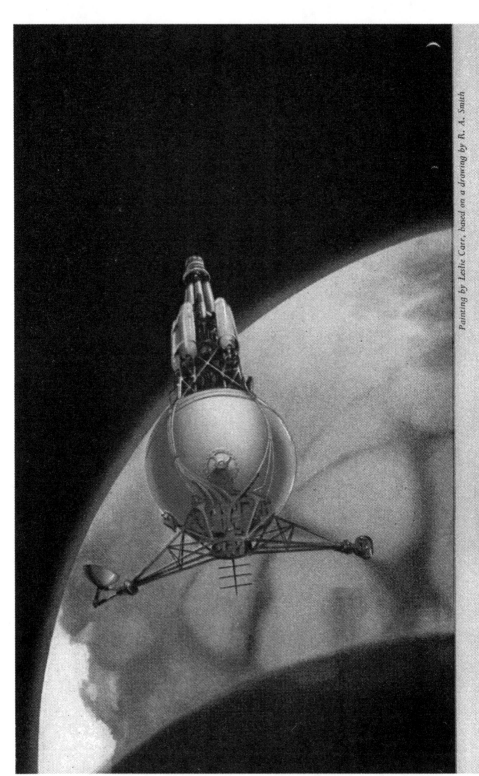

Painting by Leslie Carr, based on a drawing by R. A. Smith

AUTOMATIC ROCKET SURVEYING MARS

To Jim, who suggested it
To Fred, who provided the environment
To Dot, who had to read my writing

Contents

List of Plates

Preface

THIS book has been written to fill a need which has become increasingly apparent since my earlier work *Interplanetary Flight* was published little more than a year ago. The latter book was intended as a technical though non-specialist treatment of astronautics—the science of space-travel—but it was soon clear that it had a large sale among readers who were certainly less than enthusiastic about the details of mass-ratios, rocket fuel performances and the dynamics of orbits.

The present work has, therefore, been prepared for the benefit of all those who are interested in the "why" and "how" of astronautics yet do not wish to go into too many scientific details. I believe that there is nothing in this book that the intelligent layman could not follow: he may encounter unfamiliar ideas, but that will be owing to the very nature of the subject, and in this respect he will be no worse off than many specialists.

In this work I have also attempted to cover a considerably wider field than was possible in the earlier volume. I have tried to give concrete answers to such questions as "What would a spaceship look like?", "What may we expect to find on the planets?", and, above all, "What will we *do* when we get there?" Very obviously, any such replies must at present be based on a most meagre foundation of exact knowledge, and I have little doubt that many of them will look rather odd in the near future. But unless *some* attempt is made to deal with these points, the whole subject remains, as far as the ordinary reader is concerned, in the realm of theory. The experts may be satisfied with graphs and equations: most of us prefer more substantial fare.

I have, therefore, not been afraid to use my imagination where I thought fit: those who want a more exact, quantitative treatment can find it in *Interplanetary Flight*. Yet I have tried to base all my speculations firmly upon facts, or at least upon probabilities, and have avoided sensationalism for its own sake. Some readers may find this a little difficult to credit—particularly when they look at Plates V and IX—but it is the truth. Space-travel is a sufficiently sensational subject to require no additional embellishments, and in the long run we can be sure that our wildest flights of fancy will fall far short of the facts—as has always happened in the past history of scientific prediction.

I do not expect all my readers to accept unreservedly *everything* I suggest as a possibility of the future, but I would ask those who may find it hard to take seriously the idea of colonies on the Moon and the planets to consider this question: What would their great-grandfathers have thought if, by some miracle, they could have visited London Airport or Idlewild on a busy day and watched the Constellations and Stratocruisers coming in from all corners of the earth? This is the experience which, perhaps above all others that the modern world can give, should convince any unbiased mind that we are already far closer in time to the first ships of space than to the first ships of the sea.

In writing this book I have tried to anticipate all the questions that the reader will ask, and have also attempted to explain everything that *is* explainable in a work of this length and scope. Some matters, however, the reader will have to accept on trust. Thus he will find a straightforward—I hope—demonstration of why the rocket works in a vacuum, but the reasons for the time-contraction effects mentioned in Chapter 17 he must seek in books on Relativity.

Astronautics, perhaps even more than nuclear physics, raises questions far outside the purely technical domain. The possibility of space-flight is often admitted by those who can see no point in its accomplishment, or who ask the very reasonable question: "Why bother about the planets when there is so much to do here on Earth?" Many people today have had quite enough of science for the sake of science and look with distaste, or even active hostility, on the extension of Man's powers which is represented by interplanetary travel. An attempt has therefore been made, in the closing chapter, to deal with this viewpoint and to show how astronautics may contribute to the progress of civilisation, and the ultimate happiness of mankind.

Once again it is a pleasure to thank my colleagues in the British Interplanetary Society for their assistance during the preparation of this book. Although I take responsibility for any views put forward, I can claim originality for very few of the ideas mentioned herein, most of which have been thrashed out in discussions with other members of the B.I.S. over a period of more than fifteen years.

My particular thanks are due to R. A. Smith and Leslie Carr for their splendid work on the illustrations. Besides executing Plates I, II, III, IV, XI and XII, Mr. Smith (with his colleague H. E. Ross) is almost entirely responsible for the ideas depicted there: my own contribution is limited to the middle distance of Plate II. The frontispiece and Plates V and IX are also largely based on Mr. Smith's designs, though the execution is entirely Leslie Carr's.

Thanks are also due to the British Astronomical Association for permission to reproduce Plates VII and X from the *Memoirs* of its Lunar and Mars Sections; to Messrs. John Murray for Plate VI(*a*), taken from Nasmyth and Carpenter's classic *The Moon;* and to the Royal Astronomical Society for Plates VI(*b*) and XIV. Plate XIII is based on information from Otto Struve's *Stellar Evolution* (Princeton University Press).

Finally, I would again like to express my particular gratitude to my friend A. V. Cleaver for his careful reading of the MS. and his many valuable, and frequently pungent, suggestions and criticisms.

London. May 1951.

THE EXPLORATION OF SPACE

1. The Shaping of the Dream

Come, my friends,
'Tis not too late to seek a newer world.
To sail beyond the sunset, and the baths
Of all the western stars.

TENNYSON—*Ulysses*

THE very conception of interplanetary travel was, of course, impossible until it was realised that there *were* other planets. That discovery was much later than we, with our scientific background, sometimes imagine. Although Mercury, Venus, Mars, Jupiter and Saturn had been known from the very earliest times, to the ancients they were simply wandering stars. (The word "planet" means, in fact, "wanderer".) As to what those stars might be, that was a question to which every philosopher gave a different reply. The followers of Pythagoras, in the sixth century B.C., had made a shrewd guess at the truth when they taught that the Earth was one of the planets. But this doctrine —so obviously opposed to all the evidence of common sense— was never generally accepted and, indeed, at the time there were few arguments which could be brought forward to support it. To the ancients, therefore, the idea of interplanetary travel, in the literal sense, was not merely fantastic: it was meaningless.

However, although the stars and planets were simply dimensionless points of light, the Sun and Moon were obviously in a different class. Anyone could see that they had appreciable size, and the Moon had markings on its face which might well be interpreted as continents and seas. It was not surprising, therefore, that many of the Greek philosophers—and not only the Pythagoreans—believed that the Moon really was a world. They even made estimates of its size and distance, some of which were not far from the truth. Once this had been done, it was natural

1

to speculate about the Moon's nature, and to wonder if it had inhabitants. And it was natural—or so at least it seems to us—that men should write stories about travelling to that mysterious and romantic world.

In actual fact, only one writer of ancient times took advantage of this now classic theme. He was Lucian of Samos, who lived in the second century A.D. The hero of Lucian's inaccurately entitled *True History* was taken to the Moon in a waterspout which caught up his ship when he was sailing beyond the Pillars of Hercules—a region where, as was well known in those days, anything was likely to happen.

In a second book, Lucian's hero went to the Moon quite intentionally, by making a pair of wings, after the fashion of Icarus, and taking-off from Mount Olympus. For in Lucian's time, as for many centuries to come, it was not realised that there was a fundamental difference between aero- and astronautics. In A.D. 160 it seemed natural enough to imagine that, if one *could* make a workable pair of wings, they could be used to take one to the Moon.

After Lucian, the theme of space-travel was neglected for almost fifteen hundred years. When it was again renewed, it was in a very different intellectual climate. The modern era had begun: the Earth was no longer believed to be the centre of the universe. And, above all, the telescope had been invented.

It is hard for us to imagine astronomy as it was in the days when all observations had to be made with the naked eye. We now take the telescope for granted, but it is little more than 300 years since Galileo pointed his first crude instruments at the stars and learned secrets withheld from all other men since history began. Few scientists can ever have gathered so rich a harvest in so short a time. Within a few weeks Galileo had seen the mountains and valleys of the Moon—proving that it was indeed a solid world—and had also discovered that the planets, unlike the stars, showed visible discs. He had found that four tiny points of light revolved around Jupiter as the Moon revolves around the Earth, and the inference was obvious that Jupiter was a world with four satellites as against Earth's one, appearing small only because of its immense distance. This was the first direct revelation of the true scale of the universe: astronomers had calculated the distances of the planets before, but now at last Man had an instrument with which he could actually *see* into the depths of space. From this moment, the old medieval conception of the universe, with its picture of concentric crystalline spheres carrying the planets between Heaven and Earth, was doomed. The frontier of space receded to an enormous distance: they are receding from us still.

It is hardly surprising that the first serious story of a journey to the Moon appeared within a generation of Galileo's discoveries. It is, however, a little surprising that it was written by the greatest astronomer of the time—indeed, one of the greatest of all time. Johannes Kepler was the first man to discover the exact laws governing the movements of the planets—the same laws which will one day govern the movements of spaceships. During the latter years of his life Kepler wrote, but did not publish, his *Somnium*. In this book, he transported his hero to the Moon by supernatural means—a retrograde step, one might think, for a scientist. But Kepler lived in an age which still believed in magic, and indeed his own mother had been charged with witchcraft. He undoubtedly employed demonic methods of propulsion because he knew of no natural forces that could undertake the task. Kepler—unlike his predecessor Lucian—knew perfectly well that there was no air between the Earth and the Moon, although he thought that the Moon itself might have an atmosphere and inhabitants. His description of the Moon was the first one to be based on the new knowledge revealed by the telescope, and it had a great influence on all future writers (including H. G. Wells, two and a half centuries later).

Kepler's book was published in 1634. Only four years afterwards the first English story of a lunar trip appeared—Bishop Godwin's *Man in the Moone*. Godwin's hero, Domingo Gonsales, flew to the Moon on a flimsy raft towed by trained swans. This feat was really quite accidental, for Gonsales had merely been attempting the conquest of the air, not of space. But he did not know that his swans had the habit, hitherto unrecorded by ornithologists, of migrating to the Moon. His involuntary flight to our satellite occupied twelve days, and he had no difficulty with breathing on the way. However, he did notice the disappearance of weight as he left the Earth, and on reaching the Moon discovered that its pull was much weaker, so that one could jump to great heights. This idea is now quite familiar to us, but Godwin was writing fifty years before Newton discovered the law of gravitation.

The idea of lunar voyages was now becoming popular, and in 1640 Bishop Wilkins published a very important book *A Discourse Concerning a New World*. This was not fiction, but a serious scientific discussion of the Moon, its physical condition, and the possibility that it might have inhabitants. But Wilkins went further than this, for he concluded that there was no reason why men should not one day invent a means of transport—a "flying chariot", as he called it—which could reach the Moon. He even suggested that colonies might be planted there, a proposal which, needless to say, caused some foreign writers to make rude remarks about British imperialism.

During the next two centuries, there was a steady trickle of books about space-flight. Some were pure fantasy, but others made at least occasional attempts to be scientific. Undoubtedly the most ingenious writer during this period was Cyrano de Bergerac, author of *Voyage to the Moon and Sun* (1656). To Cyrano must go the credit for first using rocket propulsion, even though he certainly had no idea of its advantages. Still more surprising, he anticipated the ramjet. In his last attempt at interplanetary flight, he evolved a flying machine consisting of a large, light box, quite airtight except for a hole at either end, and built of burning glasses to focus sunlight into its interior. The air, being thus heated, would escape from one nozzle and be replenished through the other. Unfortunately, Cyrano was still obsessed with the notion that "nature abhors a vacuum"—it had not yet been discovered that most of nature *was* a vacuum—and he imagined that his machine would be propelled skywards by air rushing into the lower orifice instead of out of it.

Although most of the stories of this era were concerned with voyages to the Moon (and sometimes to the Sun, which was also believed to be a habitable world) some writers' imaginations did go a little further afield. Thus de Fontenelle, in 1686, wrote a widely read book on popular astronomy in which he maintained that all the planets were inhabited by beings who had become suitably adapted to their surroundings. And in 1752 the great Voltaire produced *Micromegas,* a work which is remarkably modern in outlook as it shows Man and his planet in the correct astronomical perspective. Micromegas was a giant from the solar system of Sirius, and visited Earth with a companion from Saturn. Like so many works of its kind, before and after, *Micromegas* was used chiefly as a vehicle for satire.

By the dawn of the nineteenth century, however, the space-travel story had run into trouble. Too much was known about the difficulties and objections to interplanetary flight, and science had not yet advanced far enough to suggest how they might be overcome. The invention of the balloon (in 1783) had diverted attention to travel inside the air, and had also shown conclusively that men could not live unaided at great altitudes. The Moon and planets had become much less accessible than they had seemed to Bishops Godwin and Wilkins.

By the second half of the century, however, the fiction writers had overcome their momentary embarrassment, and stories of space-travel became both more common and more scientific. No doubt the great engineering achievements of the Victorian age had produced a feeling of optimism: so much had already been accomplished that perhaps even the bridging of space was no longer a totally impossible dream.

This attitude is apparent in Jules Verne's famous story *From*

the Earth to the Moon (1865). Although much of it is written
facetiously—Verne got a good deal of fun out of caricaturing the
go-getting Americans who were so anxious to reach the Moon—
this work is important because it was the first to be based on
sound scientific principles. Verne did not take the easy way out
and invent, as so many writers before and since have done, some
mysterious method of propulsion or a substance which would
defy gravity. He knew that if a body could be projected away
from the Earth at a sufficient speed it would reach the Moon:
so he simply built an enormous gun and fired his heroes from it
in a specially equipped projectile. All the calculations, times and
velocities for the trip were worked out in detail by Verne's
brother-in-law, who was a professor of astronomy, and the projectile itself was described in minute detail. One of its most interesting features was the fact that it was fitted with rockets for
steering once it had reached space. Verne understood perfectly
well—as many people at a much later date did not—that the
rocket could function in an airless vacuum, but he never thought
of using it for the whole trip.

It is probable that Verne really believed that his space-gun
would work, though we know now that the projectile would
have been destroyed by air resistance before it left the barrel.
On the other hand, Verne can hardly have imagined that his
travellers would have survived the initial concussion, which
would have given each of them an apparent weight of several
thousand tons. No doubt he passed off this minor point with a
light laugh for the sake of the story.

Verne never landed his heroes on the Moon, perhaps because
he was unable to think of any way in which they could return
safely. Instead, they performed a circumnavigation and then
came back to Earth—which, in fact, is almost certainly what the
first lunar spaceships will do.

H. G. Wells was less scientific but more adventurous—and also
much more readable, for his *First Men in the Moon* is one of the
very few interplanetary romances which is also a work of literary

art. On the purely technical side, it marks a retrogression from
Verne, whose space-gun was at least plausible and founded on
scientific facts. To get his protagonists to the Moon, Wells invented "Cavorite", a substance which could act as a gravity
insulator. His heroes had only to climb into a sphere coated with
this useful material, and they would travel away into space. To
steer themselves towards the Moon, it was merely necessary to
open a shutter in that direction. . . .

This conception of a gravity-insulating or defying substance did
not originate with Wells, and the first person who seems to have
employed it was one J. Atterley, whose *Voyage to the Moon*
appeared in 1827. Neither Mr. Atterley nor any of his numerous
successors ever explains, so far as we are aware, how their anti-
gravitational metals manage to stay on Earth: one would have
thought that materials with such a tendency to levitation would
long ago have departed into space.

It is not difficult to show that a substance like Wells' "Cavorite" is a physical impossibility, defying fundamental laws of
nature. But the idea of anti-gravity is not in itself absurd, and we
shall return to it in Chapter 17.

Wells' book appeared in 1901, and it would be difficult to
count—let alone read—the number of works that have since
touched upon the subject of interplanetary flight. There are two
very obvious reasons for this increase. In the first case, the conquest of the air had acted as a stimulus to imagination: in the
second, the foundations of astronautics were being laid by compe-
tent scientists, and the result of their work was slowly filtering
through to the general public. The researches of Goddard (from
1914 onwards) and later of Oberth had focused attention on to
the rocket, and even before the modern era of large-scale experimental work had proved the accuracy of these men's predictions,
the rocket had been accepted as the motive power for spaceships
in the majority of stories of interplanetary travel. As we shall see
later, this prophecy is almost certain to be correct.

Nor can it be doubted that these countless stories—and not

merely those few with a carefully scientific basis—have done a great deal to bring closer the achievement of which they told. When one considers it dispassionately, it is a somewhat extraordinary situation. Even the literature of flight, which provides the closest parallel, is nothing like so extensive or so carefully worked out. The conquest of space must obviously have a fundamental appeal to human emotions for it to be so persistent a theme over such a span of time. It is a little strange, therefore, to think that in what might be called its "classic form" the space-travel story will soon disappear, for history is overtaking imagination. When the first rocket lands on our satellite, the romantic writers will have lost the Moon: but it will be a small sacrifice, for the Universe will still remain as their playground.

2. The Earth and Its Neighbours

Beneath the tides of day and night
With flame and darkness ridge
The void, as low as where this earth
Spins like a fretful midge.

D. G. ROSSETTI—*The Blessèd Damozel*

THE first difficulty one encounters
in trying to envisage interplanetary flight is that of scale. The distances involved are so enormous, so much greater than those we
meet in everyday life, that at first they are quite meaningless. However, this is something that (with practice) can be fairly easily
overcome.

There are still primitive peoples to whom a hundred miles is
an inconceivably great distance—yet there are also men who think
nothing of travelling ten thousand miles in a few days. As speeds
of transport have increased, so our sense of distance has altered.
Australia can never be as remote to us as it was to our grand-
fathers. In the same way, one's mental attitude can adapt itself
to deal with interplanetary distances, even if the mind can never
really envisage them. (And, after all, can the mind *really* envisage
a thousand miles?)

The first step in this "familiarisation procedure" is the scale
model. To begin with, let us concentrate on Earth and Moon
alone, ignoring the other planets. We will take a scale on which
a man would still be visible to the naked eye, our reduction
factor being 1,000 to 1. The Earth is now a sphere 8 miles in
diameter, and 240 miles away is another sphere, the Moon, 2
miles across. On this scale, a human being would be a little less
than a twelfth of an inch high, the speed of the fastest aircraft
would be under a mile an hour, and that of a V.2 rocket about
three and a half miles an hour. The twelfth-of-an-inch-high man

9

contemplating the gulf between Earth and Moon is thus in much the same position as an intelligent ant trying to picture the size of England.

To bring in the planets, we must alter the scale again, making the man sink far below visibility. With a reduction of a million-fold, the Earth is now 40 feet in diameter, the Moon 10 feet across and a quarter of a mile away. The Sun is 93 miles away and almost a mile across; 36 and 67 miles from it, respectively, circle Mercury and Venus. Mercury is 15 feet across, Venus 38—a little smaller than the Earth. Beyond the Earth's orbit is Mars, 20 feet in diameter and 140 miles from the Sun. It is accompanied by two tiny satellites, only about half an inch across.

Outwards from Mars is a great gulf, empty save for thousands of minor planets or "asteroids", few of which on this scale are much larger than grains of sand. We have to travel 483 miles from the Sun—340 beyond Mars—before we meet Jupiter, the largest of all the planets. In our model he would be over 400 feet in diameter, with eleven satellites ranging in size from 15 feet to a few inches across.

You may feel that our model is getting somewhat unwieldy despite our drastic reduction of a million-to-one, but we are still nowhere near the limits of the Sun's empire. There are four more planets to come—Saturn (diameter 350 feet); Uranus (150 feet); Neptune (160) feet and Pluto (20 feet). And Pluto is 3,700 miles from the Sun. . . .

This model of our Solar System shows very clearly the *emptiness* of space, and the difficulty of representing on the same scale both the sizes of the planets and the distances between them. If we reduced the Earth to the size of a table-tennis ball, its orbit would still be half a mile across, and Pluto would be ten miles from the Sun.

A pictorial attempt to show the planets, their satellites and their orbits to the correct scale is given in Figure 1. Even in the most "magnified" of the diagrams, however, it is not possible to represent the smaller satellites accurately.

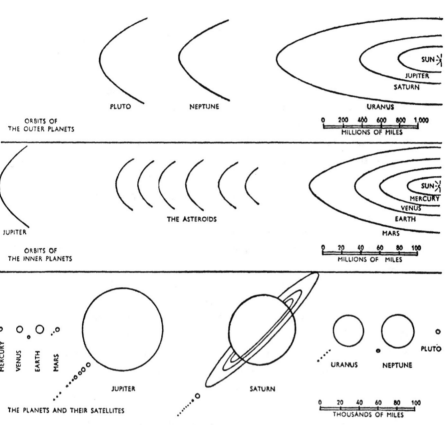

Figure 1. The Solar System

Three other points remain to be mentioned before our picture of the Solar System is complete. In the first place, it is not a stationary affair. All the planets are moving, and in the same direction round the Sun. The innermost planet, Mercury, takes only 88 days to complete one revolution, while Pluto takes 248 years—so that astronomers will have to wait until A.D. 2178 before it returns to the part of the sky where it was discovered in 1930. This increase in period from Mercury to Pluto is not merely due to the greater distances which the outer planets have to travel. They also move more slowly, for reasons which we will discuss in Chapter 4. Mercury is moving in its orbit at 107,000 m.p.h., the Earth at a more modest 68,000 and Pluto at a mere 10,000 m.p.h.

The second important point is that almost all the planets lie in or very near the same plane, so that the Solar System is virtually "flat". There are exceptions to this rule, the worst being Pluto, whose orbit is inclined at an angle of 17 degrees to that of the Earth's, but on the whole it is fairly well obeyed—and it certainly simplifies the problem of interplanetary navigation.

Finally, the shapes of the orbits. They are very nearly circular, with the Sun at the centre. Only Mercury, Mars and—once again—Pluto depart seriously from this rule, their orbits being appreciably elliptical. That of Pluto, in fact, is so eccentric that it can sometimes come closer to the Sun than does Neptune.

This, then, is the family of planets of which our world is a rather junior member. Despite its size, it forms a virtually isolated system in space, owing to the remoteness of even the nearest stars. (We shall discuss the scale of the stellar universe in Chapter 16.) With perfect precision, age after age, the planets swing in their orbits round the Sun—for they are moving in an almost total vacuum, beyond the reach of friction or any force which might check their speed.

The airlessness of space must have seemed an insuperable barrier to interplanetary voyages in the days when it was first realised that the atmosphere extends only a little way from the Earth. Today we know better—for if there was any appreciable amount of matter in space, it would not be possible to reach the speeds which are required for journeys to other worlds. However, this would be rather a theoretical consideration, since those worlds themselves would long ago have ceased to exist: the resistance to their motion would have made them spiral inwards until they dropped into the Sun.

There will be opportunity later to look more closely at the other planets, and to see what is known about their physical conditions. We must now return to Earth, which for a long time to come will be the starting point for all our voyages, and consider what obstacles we will have to overcome if we wish to leave it.

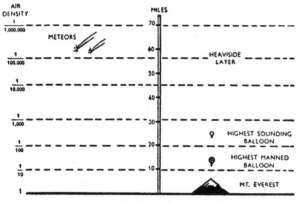

Figure 2. The Earth's Atmosphere

As we have already mentioned, the atmosphere is both a help and a hindrance. We cannot live if the pressure of the surrounding air falls below about one-half of its sea-level value, and most men would be practically incapacitated well before this figure was reached. As we ascend from the Earth's surface, the pressure and density of the air fall steadily, and in Figure 2 an attempt has been made to show this on the correct scale. The greatest height at which men can live permanently is three and a half miles, and this requires a long period of adaptation.

The absence of atmosphere does not, of course, merely affect men, but also machines. Aeroplanes—whether propelled by airscrews or jets—are air-burning devices just as much as men are, the difference being that in the human engine the combustion is gentler and the conversion of energy to power rather more subtle. In addition, aircraft require the atmosphere for support: wings and airscrews would both be useless in a vacuum. These factors set a limit to the height at which conventional aircraft can ever operate. That limit is between 10 and 15 miles—or, roughly speaking, where the air pressure is more than one-twentieth of its sea-level value.

Balloons can function at considerably greater altitudes—up to 25 miles if only carrying very light instruments—but presently they too reach a level where the surrounding air is little denser than the hydrogen in the envelope, and so can give them no more buoyancy. Until the advent of the giant rocket, no object made by man had ever risen above this level—with the single exception of the shells from the Paris gun (usually misnamed "Big Bertha") in the 1914 War. These had a peak altitude of 30 miles, and, if the gun could have been fired vertically, would have reached a height of over 40 miles.

The atmosphere, however, does not come to an end at the level where balloons will rise no further: indeed, it never comes to an end at all, but dwindles with distance as a musical note dies away with time, until at last there is no means of detecting its presence. Fifty miles up there is still enough to play a very important rôle in our modern lives, for here the tenuous ·gases are ionised to form the reflecting layer (the ionosphere or Heaviside layer) which makes long-distance radio communication possible and provides many of the strange noises which occasionally intrude upon our listening. Around this level most of the meteors which come racing into the atmosphere at speeds of 100,000 m p.h. or more meet their doom, for at these velocities even the ionosphere can produce tremendous frictional resistance.

The last indication we have of the atmosphere's presence is given by the aurora, whose ghostly beams and curtains of light we seldom see in southern latitudes, but which are common enough near the magnetic poles. The aurora is produced by electrical discharges very similar to those in neon signs, and its outermost streamers extend to heights of six hundred miles. Well below this altitude, however, we are in a vacuum better than any that can be produced in the laboratory.

It might be thought that, because the atmosphere is so relatively shallow, it could be ignored as far as interplanetary flight is con-

Plate I

HIGH-ALTITUDE MAN-
CARRYING ROCKET

Drawing by R. A. Smith

Plate II   Drawing by R. A. Smith

SPACESHIPS REFUELLING IN FREE ORBIT

cerned. This is not the case: as we shall see later, it sets a limit to the speed we can develop near the Earth's surface on the way into space and—much more important—it offers a means of making a safe landing on the return.

This thin blanket of air, without which life as we know it would be impossible, is held tightly to the Earth by the force of gravity. If gravity were weaker by a factor of four or five, the atmosphere would have escaped into space—as has happened in the case of the Moon. So we can be thankful for gravity in this respect, even though when we contemplate the task of leaving the Earth we may wish it had a much smaller value.

Of all natural forces, gravity is the most universal and it dominates any discussion of space-flight. Here on the Earth's surface we can never escape from its influence, and its value is practically constant over the whole of the planet. With increasing altitude it slowly diminishes, though so slowly that even at the greatest height yet reached by rocket (250 miles) it still has 90 per cent. of its value at sea level. As the distance from the Earth lengthens into the thousands of miles, the reduction becomes substantial: twelve thousand miles up, a one-pound weight would weigh only an ounce. It follows, therefore, that the further away one goes from the Earth the easier it is to go onwards. (A practical example of the saying that nothing succeeds like success.) As far as gravity is concerned, leaving the Earth is rather like climbing a hill which at first is very steep but later becomes more and more gentle until finally it is almost perfectly flat. Yet it is never quite flat: the Earth's gravitational pull extends throughout the universe, even if for almost all purposes it can be ignored after a distance of a million miles or so.

This picture of gravity as producing a hill, up which we have to climb to get away from the Earth, is a very useful one, and we will employ it again later. In the meantime, Figure 3 will give some visual idea of the way in which gravity falls off with distance—and should help to dispel the surprisingly common impression that it simply finishes when one is still quite close to the Earth.

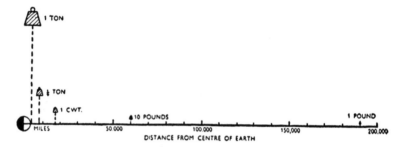

Figure 3. The Reduction of Gravity with Distance

If we ever hope to build spaceships, therefore, we must bear
in mind two fundamental points. In the first place, any method of
propulsion which depends on the atmosphere will be useless. And
secondly, even if we have a device which can produce thrust in
an airless vacuum, our ship must be provided with enough energy
to fight its way outwards for thousands of miles against the pull
of Earth's gravity.

The first condition, as we shall see in the next chapter, is easily
fulfilled. It is the second—the energy problem—which is by far
the more serious: yet even that can be solved without invoking
any new and fundamental discoveries. We do not have to wait
until someone produces "anti-gravity" before we can travel to
the planets: the means is already at hand. It is the rocket.

3. The Rocket

Oh, high aspiring, rainbowed jet!
HERMAN MELVILLE—*Moby Dick*

THE history of the rocket, though to go into it deeply would take us away from the subject of this book, is a curious and eventful one. In the field of warfare, for example, it provides a striking instance of an apparently obsolete weapon making a spectacular and perhaps final "comeback". As far as is known, it was invented by the Chinese around A.D. 1200. Although it is often said that the Chinese invented gunpowder and demonstrated their degree of civilisation by using it only in fireworks, the rocket disproves this, for they employed it against the Mongols in the siege of Kaifeng in 1232.

News of the invention reached Europe very quickly, and in the following centuries the rocket was in common use both as a firework and as an impressive but unreliable weapon. It was not of much military importance until 1805, when Sir William Congreve interested the British army in its possibilities. For a while it seemed as if the rocket might replace the gun (as indeed Congreve believed it would) but the great improvements in artillery soon made it obsolete. In the meantime, however, it had found another use—as a launcher of rescue lines to ships stranded off shore. Between 1850 and 1920, the rocket was employed almost exclusively for pyrotechnics and life-saving: both its menace and its promise still lay in the future.

During the First World War, a young American scientist named Robert Hutchings Goddard began to investigate the possible use of the rocket for the exploration of very high altitudes. In 1919 the Smithsonian Institute, which had provided financial support for Goddard's work, published his first monograph, a slim booklet of less than a hundred pages with the modest title *A Method of Reaching Extreme Altitudes*. This little book opened the modern era of rocket research, for it showed conclusively that rockets could be used to carry scientific instruments to heights never before imagined.[1] It would even be possible, Goddard concluded, to project a sufficient quantity of magnesium powder to the Moon for the flash to be observed by terrestrial telescopes.

Goddard went no further than this in discussing interplanetary travel, though it is known that his still unpublished notes (he died in 1945) contain speculations on the subject. But his influence in this field was very great, and his little book was undoubtedly a stimulus to the scientists in Europe who were beginning to look towards the planets. Most important of these was Hermann Oberth, a Rumanian professor of mathematics, who in 1923 published a monograph entitled *The Rocket in Planetary Space*. Six years later this was greatly expanded into the work which had become the "bible" of astronautics—*The Way to Space Travel*. In this remarkable book, which may one day be classed among the few that have changed the history of mankind, Oberth dealt in detail with all the fundamental problems of space-flight, not only on the purely mathematical but also on the engineering side of the subject—this at a time when the largest rocket ever built weighed only a few pounds!

The practical work on rocket engineering which took place in Germany from 1927 onwards was largely inspired by Oberth's writings, but the political background of the 1930s made it inevitable that all such research would be devoted to military· purposes. If a rocket could carry scientific instruments a hundred miles vertically, then it could carry explosives a considerably greater distance horizontally. All Goddard's initial pioneering work was financed by a grant of some $11,000, but the German

[1] Priority for the first scientific investigation of the rocket as a means of exploring space must go to Ziolkowsky (Russia) and Robert Esnault-Pelterie (France).

War Department sank £35,000,000 into the building of Peenemunde, where V.2 and many other rocket weapons were developed between 1936 and 1945. The parallel with the history of
nuclear physics is as striking as it is depressing.

The work of the German scientists and engineers at Peenemunde proved conclusively that the theories of such men as
Goddard and Oberth were basically correct. It was possible to
build large rockets, and they could attain the enormous heights
and speeds which calculation predicted. V.2 may, in many of its
essentials, be regarded as the prototype of the first spaceships.

We will say no more here about the history of the rocket,
which is unfolding every day—if the word is appropriate for
work so much of which is still secret. Instead, we will consider
how the rocket works and just why it is that it has become of
such supreme importance in astronautics.

It is often stated that rocket propulsion depends on reaction—
to put it colloquially, upon "pushing". This is perfectly true,
but it is not very helpful because it is quite impossible to think
of *any* form of propulsion which does not. Even walking depends
upon reaction: the friction of our shoes against the ground
pushes the Earth backwards and so thrusts us forwards. It was
Sir Isaac Newton who first pointed out (in his famous and
deceptively simple Third Law) that every action has an equal and
opposite reaction. The thrust we produce on the Earth is therefore exactly equal to that which the Earth produces on us, but
since there is a somewhat considerable disparity in mass between
the two bodies, we alone appear to move.

Note that in this case, it is the existence of friction that enables
a thrust to be produced. If friction were removed—if for example
we were standing on a perfectly smooth sheet of ice—walking
would be impossible.

To fix ideas more precisely, consider the following situation
(Figure 4). A man is standing on a trolley which we will assume
weighs just as much as he does. Suppose it is on a very smooth
surface, and suppose that the man jumps off it, kicking against

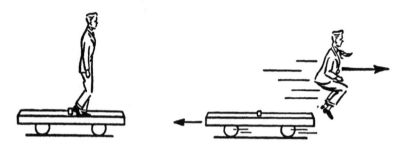

Figure 4. The Principle of Reaction

a foot-rest. Then both Newton's Third Law and common sense tell us that whatever speed the man has given himself in jumping to the right will be exactly equal to the speed with which the trolley rolls off to the left.

If on the other hand the trolley weighed twice as much as the man, its speed would be only half his, and so on for any other ratio. The law of reaction is as simple as that.

It should be noticed that in this example, unlike the case of walking, fraction did not play any part in producing the thrust. The man was able to exert the necessary "push" against some fixed part of the trolley.

A more familiar, and certainly more striking, example of reaction occurs when a gun is fired. In this case the explosion of the powder in the confined space of the breach provides the thrust, and the masses involved are very far from being equal. Yet once again the same law of reaction holds. If the gun weighed a thousand times as much as the bullet, it would recoil at a thousandth of the bullet's speed.

Now, perhaps, we can consider an analogy based on the above ideas which shows very clearly the principles of rocket propulsion. Returning to our friend on the trolley, let us imagine that his vehicle is carrying a load of bricks (Figure 5). He takes one and throws it towards the right, producing a recoil as he does so. Since the mass of the bricks is only a small fraction of that of the trolley and its cargo, the velocity with which the vehicle moves

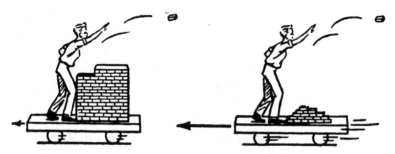

Figure 5. The Rocket Principle

off to the left is naturally very small. However, if the trolley is supported, as we assumed, on a virtually frictionless surface this velocity will not be lost.

Now the passenger throws away another brick, at the same speed as before. The trolley's velocity is immediately doubled, and as more and more bricks are hurled overboard the vehicle will steadily gain speed. It will be seen that it does not matter in the least what happens to the bricks after they have left the hand of the thrower: all the recoil or thrust is produced during the act of throwing itself. The method of propulsion is, therefore, *independent of any external medium.*

We can learn several other important lessons from this simple analogy. As the bricks are used up, so the weight of the vehicle steadily decreases. It follows, therefore, that the last bricks will produce a much greater effect than the first ones. If the "empty" weight of the trolley is half that of its "full" weight, the last brick of all will produce twice the gain in speed that the first one did. Consequently not only does the trolley's *velocity* increase during the experiment—its *acceleration* does so as well.

The analogy with the rocket should now be clear, the main difference between the two cases being that a rocket ejects matter continuously and not in separate lumps, so that it produces a steady thrust instead of a series of jerks. But the principle is exactly the same in both cases.

It is very instructive to consider how we could improve the

performance of this peculiar method of transport—our muscle-powered reaction device. Let us assume that what we are aiming at is the greatest possible final speed when all the bricks have been used up. It should be fairly obvious that this final speed depends on two factors only—(1) the speed with which the bricks are thrown out, and (2) the quantity of bricks thrown out.

Common sense, without any mathematical aid, tells us at once that doubling the speed of ejection of the bricks would double the speed of the trolley. But suppose that it was impossible to increase this any further, as would be the case if the brick-thrower were working at the limit of his strength. Could he reach a greater speed from a standing start if he took on a larger load of bricks? And in what way does the initial load affect his final speed?

The answer here is a trifle more complicated, though the main conclusions can be stated at once. The final speed *can* be increased in this manner, but it is an inefficient way of doing so. It would be far more economical to increase the speed of ejection than to increase the load of bricks, and the latter should only be done if there is no alternative.

At this stage, a few numerical values will be useful. Let us suppose that bricks are thrown off the trolley at a constant speed of 20 m.p.h., and suppose that the trolley and its energetic rider weigh 200 pounds altogether. (Once again, remember that we are ignoring friction completely.) The question we would like to answer is this—what weight of bricks will have to be taken aboard to give the trolley a final speed of 20 m.p.h., the velocity with which the bricks are ejected?

At first sight it might be thought that, since the empty weight of the vehicle is 200 pounds, the rider would have to throw off 200 pounds of bricks. This is not quite correct, for allowance must be made for the fact that extra work has to be done accelerating those bricks which are carried during the propulsion period. A simple calculation shows that when this is taken into account, the 200-pound trolley would have to take aboard a "propellant" weight, as we might call it, of 344 pounds, or 1.72 times the final, empty weight of the trolley.

Now let us be more ambitious. Is it possible for the trolley to carry so much propellant that it could finally travel *twice* as fast as the speed with which the bricks are thrown off? This idea may at first seem a little startling, but it must be remembered that as long as any propellant remains aboard, the trolley can go on accelerating, so there is no theoretical reason why it should not travel faster than the bricks themselves.

Calculation shows that the vehicle's speed could indeed be doubled, but it would mean taking on a load of bricks equal to 6.4 times the final weight of the trolley—1,280 pounds of bricks in all.

Need we stop here? In theory, no. As long as the final empty weight of the trolley and its exhausted passenger was a sufficiently small fraction of the initial weight, *any* ultimate speed could be attained. But clearly the figures are going up very steeply. For the trolley to reach three times the ejection speed, the initial weight of bricks would have to be 19 times the final weight of the vehicle, so that the trolley would have to take aboard 3,800 pounds of bricks. Yet if it were possible to triple the ejection speed, then the first example would apply once more and only 344 pounds would be needed to accomplish the same result.

This apparently far-fetched analogy has been discussed at some length because it demonstrates very clearly the fundamental laws of rocket propulsion. A rocket is, in its essentials, a device carrying a certain weight of propellant and ejecting it at the greatest possible speed, this action being accomplished by chemical energy. All the conclusions we have drawn from our example may be applied to the rocket without modification. To sum up, they are as follows:

(1) The rocket will work in the absence of air or any other medium.

(2) If the thrust or recoil remains constant, the accelera-

tion will steadily increase as propellant is used up and the rocket becomes lighter.

(3) The final speed depends directly on the ejection (or exhaust) speed: doubling the latter will double the former, and so on.

(4) The final speed also depends on the weight of fuel ejected. If this is 1.72 times the final weight of the rocket, then a speed equal to that of the exhaust will be reached. If the ratio is 6.4 to 1, the rocket can travel twice as fast as its exhaust: if 19 to 1, three times as fast, and so on.

It follows, therefore, that the rocket can be used .for the exploration of airless space, and if we are clever enough to build rocket vehicles carrying a sufficiently large weight of fuel, they can undertake any task we care to set them. Just how big that "if" is we will consider later: now, perhaps, it will be as well to look more closely at the mechanics of the rocket—to see how it is built and how it operates.

The very first rockets were nothing more than cardboard tubes packed with gunpowder, the tube being sealed at one end and having a small choke or nozzle at the other through which the hot gases escaped. Many quite large and powerful rocket motors —particularly those used for missiles and aircraft assisted-take-off—are still designed on this principle. They have the advantage of extreme simplicity, but once the propellant has been ignited there is no way of controlling it: the rocket will go on operating until all the fuel has been burned.

The newer type of rocket, of which V.2 is the most important example, is considerably more complicated but also capable of much higher performances. Instead of using solid propellants stored and burned in the same container, it uses liquid fuels which are kept in separate tanks and forced, by pumps or other means, into a carefully designed combustion chamber. Such a system, despite the additional engineering it involves, has several advantages: most important of all, it gives much higher exhaust

velocities than can be
attained with solid
or powder fuels. In
addition, liquid-pro-
pellant rockets are
fully controllable: the
thrust can be regu-
lated, just as in an
ordinary engine, by
"throttling back" the
fuel supply.

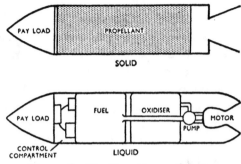

Figure 6. The Two Types of Rocket

Figure 6 shows, schematically, the two basic types of rocket. It will be seen that the liquid-propellant type must carry not only the fuel but, in addition, the oxygen to burn it with, stored in a second tank. (Solid fuel rockets must also carry their oxygen, of course, but in this case it is already combined chemically in whatever explosive is employed.) Many types of fuel have been proposed and used in practice, alcohol and petrol being the most popular. Instead of liquid oxygen, which has to be kept at 300 degrees below zero Fahrenheit to prevent it boiling away, such chemicals as nitric acid and hydrogen peroxide have been employed as oxidisers, since they contain a high proportion of oxygen.

The design of a large liquid-propellant rocket involves many engineering problems, only a few of which will be outlined here. To force the fuel and the oxidiser into the combustion chamber, turbine-driven pumps are frequently employed. In future, it is quite likely that some of the rocket exhaust itself may be "bled off" to drive these turbines.

Steering is usually effected by small vanes in the jet, operated by gyroscopes which control the path along which the rocket flies. It is also possible to steer the rocket by tilting the motor, but this requires rather complicated engineering and would certainly not be used for very large rockets.

The layout of a liquid-fuelled rocket is shown in Plate I, a design for a proposed man-carrying vehicle for high-altitude research, based on the classic V.2 conception. The pilot sits in the pressurised cabin at the nose: this is fitted with its own parachute and would be separated from the main body of the rocket during the descent. The forward tank contains alcohol, the lower one liquid oxygen, and below that are the turbine-pumps, driven by superheated steam produced by the decomposition of hydrogen peroxide—the method used in the V.2. The motor assembly, the pipe-lines for the fuel, and the control vanes in the jet are also shown.

Man-carrying rockets similar to this, or perhaps fitted with wings to enable them to glide immense distances when they return to the atmosphere, will certainly be developed during the next few decades and will be the precursors of the true "spaceships". To build them would not involve any new discoveries or fundamental advances in rocket design: the knowledge required is already available.

In the meantime, however, a great deal of useful work is being carried out with rockets containing automatic instruments which radio their readings continually back to earth during the flight. This technique of "telemetering", which does not involve danger to any human pilot, will play a very important rôle in the conquest of space, since almost certainly the first spaceships will be purely automatic—and thus expendable.

The "exhaust speed" of modern liquid-fuelled rockets is of the order of 5,000 m.p.h. or even higher. Reverting to our earlier example of the trolley-load of bricks, it follows that a speed of 5,000 m.p.h. could be reached by a rocket which weighed, say, one ton when it was empty, and could carry 1.72 tons of fuel. Such a ratio would be quite easy to achieve, and in fact V.2 does considerably better than this. However, in actual performance there are two important factors which we neglected in our previous discussion. One is friction—in this case, air resistance. This would not apply to a rocket operating in space, but it plays an appreciable part in the case of a rocket climbing through the atmosphere, cutting down its speed by five or ten per cent. (and very much more in the case of small rockets).

A much more serious factor is gravity. That did not come into the picture at all when we were considering the behaviour of the trolley on a horizontal surface. But rockets usually travel vertically, not horizontally, and this means that gravity is continuously pulling them back and reducing their speed. The effect of this reduction, for a rocket rising vertically, is 20 m.p.h. for every second of climb—or 1,200 m.p.h. for every minute.

These two factors may reduce a rocket's final velocity to about 70 per cent. of its theoretical value, but in the case of large spaceships there are ways in which this loss can be minimised—and, in certain special cases, eliminated completely.

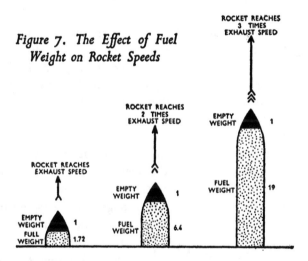

Figure 7. The Effect of Fuel Weight on Rocket Speeds

Figure 7 is an attempt to show in diagrammatic form the effect of a rocket's fuel-weight to empty-weight ratio on its performance. In each case the empty-weight has been assumed to be the same—one ton. The first rocket could be built easily: the second represents a figure which has not yet been achieved. V.2 lies between these extremes, though nearer the lower value: it carried

2 tons of fuel to every ton of empty-weight. It is probable that the second rocket could be built, which means that final speeds of twice the exhaust speed may be practicable.

But the third rocket represents something which any engineer would shudder to contemplate, since it involves squeezing 19 tons of fuel into a structure weighing only one ton. That one ton, remember, must cover the weight of the tanks, the rocket motor, the controls, the shell and framework—and the payload! It is very doubtful indeed if this could be achieved: even if it could, the payload would certainly be negligible.

We may therefore draw the important conclusion that with rockets of this type (for reasons which will be clear later we will call them "single-stage" rockets) *speeds equal to twice that of the exhaust may be attainable, but speeds three times as great are probably beyond realisation.*

Since present-day fuels give exhaust speeds of some 5,000 m.p.h., this means that they may ultimately provide us with rocket speeds of some 10,000 m.p.h. To go beyond this figure we must either develop more powerful fuels, or evolve new methods of construction. Both these lines of attack will certainly be followed up.

Before leaving the rocket for the moment, it may be advisable to look a little more closely at the operation of the motor to see exactly where and how the thrust is developed. We do this because experience has shown that, though many people can see

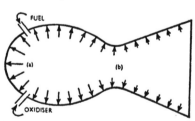

Figure 8. The Forces in the Rocket Motor

from the example of the trolley load of bricks that the rocket *ought* to work in a vacuum, they are liable to acquire gnawing doubts when they try to visualise just what happens as the hot gases rush out into nothingness.

There is nowhere, it seems, where any "push" can be produced.

The answer to this very reasonable question is given in Figure 8, which is a simplified cross-section of a rocket motor. The fuel is injected from the left and burns in the combustion chamber, releasing an enormous amount of heat and so expanding whatever gases may be produced by the reaction. As a result, a pressure is produced which acts on the chamber walls in the manner shown by the arrows. If there was no nozzle through which gas could escape, the pressures (assuming that the chamber did not burst!) would equalise and so there would be no tendency to move. However, because of the nozzle the opposing forces at (a) and (b) no longer balance, and the resultant thrust at (a) drives the rocket towards the left.

It cannot be overemphasised that the only effect of an atmosphere round the rocket would be to *reduce* its efficiency. In fact, both theory and experiment show that as a rocket leaves the atmosphere, the thrust of the motor increases by from 10 to 15 per cent. over its sea-level value.

In passing, it may also be remarked that a rocket would get no additional thrust if, when it started, the jet impinged on the ground or some other fixed obstacle. This would be undesirable in any case, as the jet might be "reflected" and could easily cause severe damage to the rocket structure itself.

With this picture of the rocket and its method of operation before us, we can now go on to discuss the magnitude of the task confronting it, if it is to be used to lift us away from the Earth.

4. Escaping from Earth

"Now, *here*, you see, it takes all the running *you* can do,
to keep in the same place. If you want to get somewhere
else, you must run at least twice as fast as that!"

LEWIS CARROLL—*Through the Looking-Glass*

GRAVITY, like the air we breathe, is one of those natural phenomena we take for granted and never think about in the ordinary course of events. It is a major factor in the lives of steeplejacks and mountaineers, but those of us who live more two-dimensional existences usually notice it only when we run upstairs in a hurry or sit on a chair which has unaccountably removed itself.

There is nothing that anyone has ever been able to *do* about gravity: it acts on all bodies in a precisely identical manner, giving any unsupported weight the same acceleration towards the Earth—an acceleration of 32 feet per second. This means that a body starting from rest would, in the absence of air resistance, be travelling at 20 m.p.h. after one second of fall, 40 m.p.h. after two seconds, 60 m.p.h. after three seconds, and so on. This value of "g" is almost constant over the whole Earth: other planets, as we shall see later, have different gravities, most of them considerably less than that of our world. On some very tiny "planetoids", gravity is so small that it would take minutes for a falling body to descend a couple of yards.

Leaving the Earth implies moving upwards against gravity, and this requires work. The amount of work involved in climbing to any given height can be easily calculated, and obviously a rocket which is designed to reach this altitude must carry enough fuel to provide the necessary work. Nature never gives us something for nothing: in fact she usually takes more than she gives back.

As has been explained on page 15, gravity steadily weakens as we go outwards away from the Earth, until at very great distances it becomes completely negligible. The analogy of climbing up a slope which is first very steep and then more gentle has already been mentioned, and it is very helpful to take this picture further: we can learn a great deal from it.

When one calculates the amount of work that has to be done in lifting a body from the surface of the Earth to a point so distant that gravity is negligible, one obtains a surprisingly simple result. The work involved is exactly the same as that needed to climb vertically through a distance equal to the radius of the Earth— say 4,000 miles—in a steady gravity field, if "g" remained constant at its sea-level value.

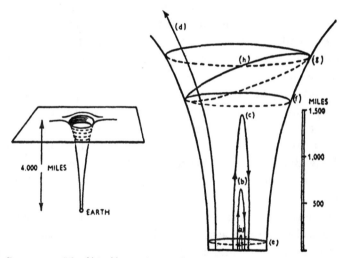

Figure 9. The "Pit" Analogy of the Earth's Gravitational Field

We dwellers on the Earth's surface, therefore, as we look up at the planets and wonder how we can reach them, are in rather the same position as people at the bottom of a perfectly smooth pit or funnel 4,000 miles deep, set in the surface of an endless, flat plain. Before we can think of travelling to other worlds, we have to climb out of this pit. The horizontal plain represents gravitationless space which, once we have reached it, we could then navigate for ever with no further expense of energy. (This is an oversimplification, for we have ignored the gravity fields of the Sun and the other planets: but for the moment we can forget these. In any case, they are quite unimportant if we are only considering the journey from Earth to Moon.)

This "gravitational pit" is shown in cross-section in Figure 9, and an enlarged view of what might be called its lower slopes is given on the right of the diagram. Imagine that it is made of some smooth, hard material like perfectly frictionless glass, and consider what would happen if a body—say a marble—were propelled up its slope at a certain initial speed. Quite clearly, it would rise until its velocity had been reduced to zero, when it would start to fall back again, eventually returning to its starting level at the speed with which it began. The heights which would be reached by bodies starting at (a) 5,000, (b) 10,000 and (c) 15,000 m.p.h. are shown in the figure.

It is obvious that there will be one critical speed at which the body will never come back, but will creep over the top of the crater, as it were, and reach the horizontal level (d). *This velocity is 25,000 m.p.h. and is called "the velocity of escape".* If a body started upwards with *more* than this critical speed, it would still retain the excess as a bonus when it escaped from the pit.

This model describes accurately the behaviour of a rocket shot vertically upwards away from the Earth, but it also throws light on another very important effect.

Most people have seen pictures of the "Wall of Death", a popular circus side-show in which a motor-cyclist drives round the inside of an almost vertical cylinder, centrifugal force allowing him to defy gravity. It will be seen that exactly the same sort of thing can happen in our model. A body set moving horizontally at any point on the "crater" wall would continue to circle indefinitely if its initial speed were sufficient. The farther from the bottom, and the gentler the slope, the more slowly the body need move to preserve its position. *This is precisely how the Moon maintains itself in its orbit as it circles the Earth.* And in just the same way the planets revolve round the Sun, prisoners of its gravitational field yet always maintaining their distance.

The speed necessary to move in such an orbit near the bottom of the crater (i.e. near the Earth's surface (e)) is 18,000 m.p.h., and this is called "circular velocity". Other similar orbits are shown at (f) and (g).

These are by no means the only possibilities that exist. Look, for example, at orbit (h). This is the path of a body which was projected horizontally at (g) but with insufficient speed to maintain itself. It fell downwards, gaining speed as it did so, until it had picked up enough speed to climb upwards again and retrace its orbit. Its path is thus not a circle, but an ellipse.

All these cases could be easily demonstrated by flicking a marble around the inside of a suitably shaped funnel: indeed, an astronomer of our acquaintance discovered this useful analogy by accident one day when he tossed a tennis ball into a large Chinese vase!

We can sum up these results, then, as follows: a body projected away from the Earth will escape from it completely if its speed is over 25,000 m.p.h. If its speed is less than this, it may either fall back again or, if it is aimed in the correct direction, it may continue to orbit the Earth indefinitely in a circular or elliptical path. For the circular orbit nearest to the Earth, the necessary speed is 18,000 m.p.h. A rocket which attained this velocity in horizontal flight, if it was just outside the atmosphere and so beyond the reach of air resistance, would circle the world for ever like a second moon. This extremely important fact is one to which we shall return when we discuss "space-stations" and the possibility of "orbital refuelling".

At this point someone may very well remark "I appreciate that a rocket could escape from the Earth, without using any more power, once it had reached this critical speed of 25,000 m.p.h.

But couldn't it do the same thing in a more leisurely fashion, travelling relatively slowly but keeping its motors running?"

The answer is yes—in theory. But glance again at Figure 9 to see what it would mean. In a "powered climb" up the frictionless slope of our 4,000-miles-deep trap, we would be fighting gravity all the way. *Even to stay still* at a fixed distance would mean continuous use of the rockets, balancing their thrust against gravity. (In this way, a rocket could use all its fuel and never get more than a few feet above the ground.) So it cannot be done, at least until we possess sources of energy immensely more powerful than any we have today. The only practicable way of escaping from the Earth is to build up escape velocity as quickly as possible— and then to cut the rockets and "coast". A commonplace but accurate analogy is given by the cyclist who builds up enough speed on the level to take him over a hill. To surmount our imaginary, yet in a sense very real, 4,000-miles-high hill we need this starting speed of 25,000 m.p.h. When we have reached it, we can relax: the Earth will never be able to hold us back.

Twenty-five thousand m.p.h.! It is, of course, a speed far greater than anything ever achieved by man or indeed by any man-made object. But it no longer seems as fantastic as it did even a decade ago. In 1940, no rocket had ever travelled more than a thousand miles an hour, but by 1950 the record was five times this—and rockets had grown in size out of all recognition. Can this advance be expected to continue—will the further five-fold increase in speed we require be reached by 1960?

To answer this question, we must revert to our earlier discussion of rocket principles (page 19), in which we stated that it was impracticable to build rockets that would travel at more than twice their exhaust speed, because of the amount of fuel required. This set a speed limit—if present-day fuels were employed—in the region of 10,000 m.p.h., or considerably less than half the velocity needed for escape from Earth.

We are still, of course, in the early days of rocket development, and much more powerful fuels will be employed by the rockets of the future. Many are already known, in fact, but we cannot yet build the motors to harness them safely. It is obvious, therefore, that exhaust speeds and hence rocket speeds will continue to increase.

However, there is a definite limit to this process—a "ceiling" to the performance we can expect by the use of chemical fuels, even when they have been developed to the utmost. That ceiling appears to be at about double the values we can achieve today, and some experts would put it considerably lower. Even taking the optimistic view, therefore, it appears that the maximum speed we can ever expect from single-stage, chemically fuelled rockets is not more than 20,000 m.p.h.—and it will take many years of development to reach this figure. When allowance is made for air resistance and other losses, such rockets might just manage to enter circular orbits round the Earth, *but they could not achieve escape velocity.*

At this stage the disappointed reader may well wonder indignantly what all the fuss has been about, since interplanetary travel is now proved to be impossible. Before proceeding any further, however, it is a good idea to recall how many other things have been proved impossible in the past. One hundred and fifty years ago, distinguished mathematicians demonstrated that the new-fangled steamships could never cross the Atlantic— the coal consumption would be far too great. A century later, the birth of flight was preceded by a similar chorus of Cassandras —most famous of whom was the great American astronomer Simon Newcomb. (In 1903 Newcomb wrote a paper which should have convinced the Wrights, had they read it, that they were wasting their time. Even five years later the good professor was still stoutly maintaining that an aeroplane could never carry the weight of a passenger as well as a pilot!)

The lesson of this is that it is never safe to make what may be called "negative" prophecies in the field of scientific and technical achievement. Even if all the known facts are correctly marshalled—a much more difficult feat than may be imagined—

and a certain line of procedure is shown to be unprofitable, alternative solutions usually turn up in time.

In this case, one of the alternatives is the multi-stage or "step" rocket. It will be obvious that if we build a rocket capable of carrying a certain payload, and make *that* payload another rocket carrying the same percentage of fuel, then when the smaller machine has burned its fuel it will have achieved twice the speed that either rocket could do by itself. Moreover, this process may be repeated: there is no need to stop at two stages. If we build a rocket with a sufficient number of steps, *any desired terminal speed may be achieved.*

It will, of course, be achieved at a considerable cost. The payload of a high-performance rocket is not likely to be more than a twentieth of its total weight, which means that each step will be about twenty times the weight of the steps above. For a one-ton final payload, therefore, the total starting weight is likely to go up in this sort of ratio: Single-step: 20 tons; Two-step: 400 tons; Three-step: 8,000 tons. And so on!

It may be asked why a three-step rocket of 8,000 tons initial weight would perform better than a single-stage one of the same weight. The answer is that, in this extremely simplified example, the single-stage rocket could carry a payload of 400 tons and give it a certain speed, which could not be greatly increased even if we cut the payload down to one ton and made the remaining 399 tons into fuel. The three-step rocket, on the other hand, could triple the speed of the one-ton payload. It is an old story—one has to pay for speed. And speed is the first essential in the problem of space-flight.

A striking example of the step principle in operation is provided, somewhat unexpectedly, in the early history of polar exploration. If one man, by his own efforts, could only carry enough provisions for a journey of a hundred miles, it is possible to extend this range by starting with a large number of men who turn back at some point, after handing on their surplus stores to a smaller party. If necessary, this group may repeat the procedure at a later stage. It is not an efficient method when one considers the amount of material and effort needed to produce a relatively small result. But it is a way, and at the time it was the only way, in which the limitations set by human strength and endurance could be overcome. In the same manner, we may regard the step principle as a means of overcoming the limitations of the rocket.

An example of a two-step rocket which, though it was never built, was investigated in some detail is the wartime German project known as A.9/A.10. The upper step (A.9) was a winged and possibly piloted version of V.2. It was to be lifted to a height of 16 miles and given a starting speed of over 2,500 m.p.h. by A.10, a 68-ton "booster" which would return to earth by parachute after its work was done. The little A.9 would go on to reach a speed of 6,000 miles an hour and on returning to the atmosphere would enter a high-speed glide which would take it 3,000 miles in 45 minutes.

During the Second World War, the Germans developed and used operationally, on a small scale, an interesting four-step rocket known as "Rheinbote". It was propelled by solid fuels and had a range of about a hundred miles. More recently, two-step rockets have been launched in the United States, using a V.2 for the lower stage and a small American rocket ("WAC Corporal") as the upper component.

Many papers and calculations have been published during the past few years showing how, by the use of existing fuels and materials, it would be possible to build rockets which could escape from the Earth carrying a small payload of automatic instruments. Ingenious methods of design have been evolved to extract the last ounce of advantage from the step principle and to conform with what might be called the prime law of spaceship economics: *"No dead weight shall be carried for a moment longer than necessary."*

It is, for example, obviously wasteful of energy to go on lifting the weight of large fuel tanks when they are nearly empty. If the tanks could be built in fairly small units, and dropped as soon as their contents had been exhausted, there might be a considerable gain in efficiency—though at the expense of structural complications. This method of design has been called "expendable tank construction" and a number of projects based on it have been described. As an example, Gatland, Kunesch and Dixon of the British Interplanetary Society have shown how, by the use of existing V.2 motors, a payload of 110 pounds could be sent into an orbit round the Earth. The rocket would be a three-step one with a total weight of 150 tons, the first step being powered by a group of seven V.2 motors. (In addition, a large number of solid-fuel booster rockets would be used at take-off.) The fuel for this stage would be carried in two large nose-tanks (the first to be jettisoned) and a set of annular tanks would form the body of the rocket, inside which the second and third steps would nestle.

One hundred and fifty tons of rocket and fuel to place a mere 110 pounds of instruments in an orbit close to the Earth appears a very large effort for a small return, but it should be pointed out that this design assumed only the use of existing fuels and motors. With the improvements in rocket performance which the future will bring, much smaller missiles would be able to perform the same task. For example, if hydrogen were used as fuel instead of alcohol, a rocket weighing about 30 tons might achieve orbital velocity with this 110-pound payload—and a 50-ton rocket could escape from the Earth completely.

The step rocket, therefore, already gives us the power to leave the Earth by proxy, if not in person. The initial stage of interplanetary exploration will probably be carried out in this way, for the construction of unmanned missiles, though it will involve very difficult engineering problems, will be far less costly than the construction of crew-carrying "spaceships". This is not merely because of the bigger payloads that will obviously have to be employed, but for much more fundamental reasons. A guided missile can be shot out into space, can radio back information until it is out of range or its electrical power has been exhausted, and can then be "written off". But a spaceship has to carry fuel for the return journey and any intermediate landings and take-offs. When one allows for this, the initial weight of a chemically fuelled spaceship on taking off from the Earth would be not hundreds but hundreds of *thousands* of tons—and the whole project becomes, if not impossible, certainly fantastic. We will see later how this difficulty may be avoided: for the moment let us explore the possibilities of purely automatic rockets—which might be called "reconnaissance missiles".

A great deal of scientific equipment has already been developed for use in V.2's and other high-altitude rockets. Some of the earliest experiments consisted simply of mounting movie cameras in suitable fairings round the base of the missile and so obtaining some of the most dramatic films ever made—showing the Earth dropping away as the rocket climbed into space. Before long, undoubtedly, television equipment will also be carried in this way so that the problem of retrieving the film will not arise.

These pictorial records, although spectacular and exciting (and also of some meteorological interest), are much less important than the information obtained by the instruments carried in the rocket's payload. Now that, for the first time, it has become possible to send equipment into these hitherto inaccessible regions, a whole technique of "micro-instrumentation" and "telemetering" is growing up. The problem is to design small, light instruments which can transmit their readings on a radio carrier wave so that, during the flight of the rocket, continuous records can be made at ground stations. This means that there is no need to recover the instruments—a possibility which would not arise in any case in the missions we are considering here. Whatever happens to the rocket, all the information obtained will be safely gathered in by cameras and recorders on the ground.

Some of the radio equipment already used in V.2's can transmit the readings of as many as fifty separate instruments, and the frequency used is so high that the ionosphere cannot act as a barrier. Because of the limited heights so far reached by rockets, the ranges required have been only two hundred miles or so, but interplanetary ranges would be available with special techniques. (Remember that we have already obtained radio *echoes* from the Moon—a feat hundreds of times more difficult than sending a one-way signal over the same distance!)

During the next decade it would certainly be possible to build an orbiting rocket carrying a small television camera and instruments which could radio back to earth measurements of ultra-violet and cosmic radiation intensity outside the atmosphere, and similar information which is of extreme scientific and technical importance. Such new knowledge would produce great advances in weather forecasting and radio communication, to mention only two subjects.

It is also obvious that such an orbital missile would have very great military value—which is, however unfortunate this may be, a fact assuring its early development. One cannot help wondering just how the statesmen and international lawyers will react to the situation when rockets begin to range out in world-circling orbits with a complete indifference to the geography beneath them. The extent to which any nation can claim the ownership of space in a vertical direction must clearly have some limit: once that limit has been agreed, a country would appear to have no redress if an inquisitive neighbour started making rocket reconnaissances as long as they were at a legal height!

We will return, in Chapter 15, to some of the other feats which orbital rockets will make possible, and will touch now on a considerably more exciting idea—that of sending an automatic missile to the Moon. If a rocket achieved escape velocity, and was aimed in the correct direction, it would reach the Moon rather less than five days after leaving the Earth. Remembering our picture of the Earth's gravitational field (Figure

9) we see that once it had built up a speed of 25,000 m.p.h. (which would take only a few minutes of powered flight) the rocket would "coast" upwards, steadily losing speed, until at a great distance it was moving away from the Earth at a speed of only a few hundred miles an hour. Now the Moon has, of course, its own gravitational field, which may be regarded as a much smaller version of the Earth's. Returning to our analogy on page 31, it is as though some distance away from the 4,000-miles-deep pit at the bottom of which we live there was another pit, this time only 170 miles deep. Dropping into this pit as it approached the Moon (or in more conventional terms, falling in the Moon's gravitational field) the rocket would gain speed and, at the moment of impact, would be travelling at over 5,000 m.p.h.

It will be seen, however, that if it was not aimed directly at the Moon but a little to one side, and if some judicious rocketing was employed, the missile could get into an orbit around the Moon and would continue to circle it indefinitely. It could then televise back to Earth not only close-ups of our satellite's visible face, but could also give us our first view of the hidden regions on its far side.

Nor need the Moon be our only target, by any means. As will be shown in the next chapter, once one has escaped from the Earth the journeys to the planets, though the distances involved may be hundreds of millions of miles, require very little *extra* energy. A striking example of this is given by the fact that though an initial speed of 25,000 m.p.h. is needed for a rocket to reach the Moon (closest distance 240,000 miles) a rocket launched in the correct direction at 26,000 m.p.h. would reach Venus (closest distance 26,000,000 miles)! Practically the same speed would enable Mars to be reached (closest distance 35,000,000 miles). Once again, if the technique of automatic control has been sufficiently highly developed, it should be possible to steer the rocket into an orbit round the planet and to send back information to Earth.

An impression of such a missile is given in the frontispiece. The little rocket (the last step of a far larger machine) left the Earth 250 days ago and during that time has been coasting freely, like a comet, along the path that leads to Mars with the least expenditure of fuel. It has now exhausted its last reserves, changing its orbit into one which will make it circle Mars for ever. Under the guidance of a tiny yet extremely complex electronic brain, the missile is now surveying the planet at close quarters. A camera is photographing the landscape below, and the resulting pictures are being transmitted to the distant Earth along a narrow radio beam. It is unlikely that true television would be possible, with an apparatus as small as this, over such ranges. The best that could be expected is that still-pictures could be transmitted at intervals of a few minutes, which would be quite adequate for most purposes. In addition spectroscopic measurements, magnetic observations, and many other types of record would be radioed back to Earth. At a distance of a thousand miles from the surface, the rocket would circle Mars once every five hours, so that in that time the planet would appear to change from new to full and back again.

This missile will certainly look peculiar to anyone who imagines that rockets must be sleek, streamlined projectiles with sharply pointed noses. But such refinements are not only unnecessary but actually wasteful on rockets which are launched in airless space. This reconnaissance missile would be carried inside a much larger rocket on its way up through the Earth's atmosphere, its outrigger arms possibly folded together during this stage and extending when it had entered space.

The construction of such an exploring rocket would involve exceedingly difficult problems in electronics and communications, but there is nothing about it which would be impossible. It seems more than likely, therefore, that before we begin our own travels round the Solar System, we will send these scouts ahead of us to blaze the trail and report on the conditions we may expect to meet.

5. The Road to the Planets

Are thy wings plumed indeed for such far flights?
WALT WHITMAN—*Passage to India*

IN the last chapter, we touched on the subject of interplanetary orbits—as opposed to "mere" circumnavigations of the Moon—and the somewhat surprising statement was made that, under favourable conditions, the planets could be reached with little more trouble than the Moon itself. Before we go on to discuss the building of true, man-carrying spaceships, we must look into this matter more closely.

When we are contemplating a voyage from the Earth to the Moon, these bodies are—astronomically speaking—so close together that everything else in space may be ignored. The picture is different when we consider a journey from Earth to any of the planets, for now we must take into account the presence of the Sun.

The reason is quite simple. It is the Sun's enormous gravitational field that holds all the planets, near and far, circling in their orbits as described in Chapter 2. We can visualise the Sun's influence as producing a "gravitational pit" like the Earth; but immensely deeper—on the scale we used before, not 4,000 but 12,000,000 miles deep! Like a group of motor-cyclists racing round one above the other on the sides of a giant "Wall of Death", the planets circle at their various distances. To move from an outer planet to an inner one means first of all losing speed slightly to drop down towards the Sun, then slowing down a little more as one goes past its orbit to match one's velocity with the inner planet. In exactly the same way, going to an outer planet means first *increasing* speed to climb upwards, then adding on a little more speed when passing the outer planet in order

43

to keep up with it and avoid falling back again. These manœuvres, and the velocities required for them, in the case of
Mars and Venus are shown in Figure 10. The changes in speed
provide the energy needed to move up and down the "slope" of
the Sun's gravitational field.

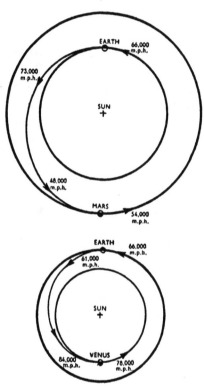

It is extremely fortunate
for astronautics that, at the
distances of the planets
from the Sun, the "slope"
of this field is relatively
gentle, and so little energy
is needed to move along
it. In most cases, in fact,
much more energy is re-
quired in the escape from a
planet's own gravitational
field—something which has
to be done in the first thousand miles or so of the
voyage—than in the jour-
ney from orbit to orbit,
which may cover a distance
ten thousand times as great.

Once again it should be
emphasised that, to travel
from one planet to another,
rocket power need only be
used for a few minutes at
the beginning and end of
the journey. If the initial

*Figure 10. The Easiest Routes
to Mars and Venus*

velocity has been correctly given, the far longer period of "free
coasting" will follow automatically.

In this sense, there is an analogy between a spaceship and an
artillery shell. Once the shell has been launched in the right
direction, no more work need be done, even though it may still

have a great distance to travel. It is much the same with the spaceship—with the important difference that whereas the shell acquires its velocity in a few feet under an enormous acceleration, the spaceship has thousands of miles in which to build up speed at a rate which human passengers could withstand. Still, this distance is as negligible compared with the total journey as the length of the gun barrel is compared with the trajectory of the shell, so when we are considering interplanetary journeys we can forget about the initial period of acceleration and imagine that the spaceship starts with the required velocity.

The orbits shown in Figure 10 are, however, by no means the only paths that would take us from Earth to Mars and Venus: they are merely the most economical ones. Faster and more direct routes are conceivable, but they would require very much greater expense of energy. With unlimited power, it would be possible to travel from one planet to another in virtually a straight line, but this will remain a dream until long after astronautics has become a fully established science. The reason why the routes shown in Figure 10 are the easiest is not hard to see: the rocket has to "take off" and "land", if one can extend these phrases to cover such a case, in the directions in which the planets are already moving. The shorter paths would cut across the planetary orbits at an angle, and so it would require much larger velocity changes to match speeds.

One important—indeed, fundamental—point about these orbits is that one could not take off for Mars or Venus at any time one pleased. The moment of departure would have to be calculated so that, when the spaceship reached the planet's orbit, the planet was there too! In the case of the Venus journey, which would last 145 days, Venus would have made almost three-quarters of a revolution round the Sun while the ship was travelling to meet her. If the appointment was not correctly made, or the ship had run out of fuel and so was unable to match speeds, it would go past Venus and would swing back to the Earth's orbit, completing the elipse and returning to its original point 290 days after departure. Unfortunately for the occupants, if they had survived, the Earth would now be a long way away, so they would have no chance of being rescued. In fact, the spaceship would have to make *five* complete circuits round the Sun, taking four years in the process, before it came near Earth again!

Missing an appointment may or may not be a serious matter in everyday life, but in interplanetary travel it would, almost inevitably, be fatal.

We can now sum-up the requirements for any interplanetary voyage, as follows:

(1) The ship must build up sufficient speed to escape from the Earth, and when it has done this must still have enough velocity left to put it into the required "voyage orbit". This means that it must start with a speed greater than that needed merely to escape—though in most cases only slightly greater.

(2) When it approaches the planet of destination, it must use rocket power again to match the speed of the planet, *and* to lower itself safely into the planet's gravitational field.

It will be seen that this is rather more than we had to do when we wanted to send a guided missile to survey another planet. This time we have the additional complication of the landing. And we have not yet discussed the return journey!

It is this latter factor that makes the problem of interplanetary flight so exceedingly difficult. Although there are people who, for the sake of adventure or scientific knowledge, would un- doubtedly undertake one-way trips (particularly if there was a chance of survival when they had landed), one can hardly make serious plans on this basis. Spaceships will have to carry enough fuel for the round trip, which involves the same problems, and exactly the same velocity changes, as the outward journey.

It can be said at once that there is no possibility of fulfilling these requirements, for even the easiest of interplanetary return journeys, by the use of chemically fuelled step-rockets alone. The starting weights would be enormous—millions of tons, in fact. Does this mean that, even if we can send television cameras to the other planets, we will never be able to go there ourselves? Or must we wait until some completely new and very much more powerful fuel is discovered—something which, in Chapter 3, we said was most unlikely?

There is an answer to this problem: indeed, there are at least two answers—one a probability, the other a certainty. Although we may never be able to increase rocket performances very greatly by the use of *chemical* fuels, there are reasons for believing that, in time, we will be able to harness atomic energy for rocket propulsion. This would certainly solve our difficulties, but it would be unwise to count on it. The other alternative, which involves no new inventions or hypothetical discoveries, is based on an idea which, though simple, at first sight seems too fantastic to be taken seriously. If we cannot build spaceships to make round trips to the planets in a single operation, then we will break the job down into a number of separate stages, *refuelling the ship when necessary.*

No motor-car, however large its tanks, could drive round the world on a single filling—but any car could do so if it made enough stops. It may be objected that there are no filling stations in space. True enough: but there is no reason why there should not be.

"Orbital refuelling", as it has been christened, is the key to interplanetary flight. It depends simply on the fact that once a spaceship had reached circular velocity outside the atmosphere, it would continue to orbit indefinitely without the use of power. Other rockets could then climb up from Earth into the same orbit and—because they would then be travelling at the same speed—could transfer fuel until the first machine's tanks had been replenished. It would then be in a position to break away from the Earth and travel out into space. Because it already possesses an orbiting speed of 18,000 m.p.h., and escape velocity is 25,000 m.p.h., the additional speed the ship would require to leave the Earth completely is thus only 7,000 m.p.h.

Before we discuss this procedure at any more length, however, it may be as well to look at the basic problems involved in the building of spaceships for simpler missions, such as entry into a circular orbit around the Earth. Once we have shown that this can be achieved, the more difficult feats of interplanetary navigation will only require extensions of the same technique.

What may be regarded as an embryo spaceship was shown in Plate I, which we mentioned on page 26. This project was designed to reach a speed, after fuel combustion, of 5,000 m.p.h., which would have taken it to a height of 225 miles. The little cabin contained only the pilot and the equipment needed for his safety during the flight. (What this equipment would have to do we will discuss in Chapter 6.)

Single-stage, manned rockets of this type, using the fuels and motors we may *ultimately* develop, could be built to reach speeds of perhaps 15,000 m.p.h. and heights of 2,000 miles, but they would be unable to reach circular velocity and so would have to return to Earth.

To get into an orbit we should need something like the two-step rocket mentioned on page 37, though even with this design much more powerful fuels than any we have today would be required. It could be done, however, and the little winged rocket could be established as an artificial satellite circling the Earth a few hundred miles up, completing one revolution every ninety minutes.

But how could the pilot get back again? To *destroy* speed by rocket power alone requires just as much effort as to attain it, and as long as the spaceship possesses its velocity it can never leave its orbit. It can no more fall down than can the Moon.

This is where the Earth's atmosphere—which was a nuisance on the way up—proves to be an advantage. Suppose the ship still has a little fuel, and suppose that by some means (see page 74) it is turned round in space so that the rocket motor points in the direction of motion. Firing the rockets will then reduce the ship's speed and it will fall towards the Earth. Quite a small change in speed will alter its path from a circular one into an ellipse *grazing the atmosphere*. As it cuts through the upper atmosphere, air resistance will reduce its speed still further and, in the process, it will become hot owing to friction. However, if the entry into the atmosphere were made at the correct angle, there would be no danger of the ship becoming incandescent: it would have nothing like the speed which brings meteors to such a spectacular end.

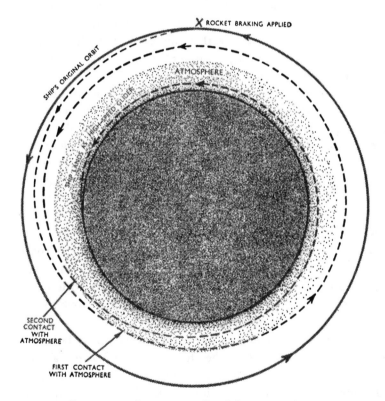

Figure 11. Returning to Earth by Air-Braking

On its first contact, the ship would stay inside the atmosphere for only a few minutes before whipping out into space again. Because of the loss in speed, however, the next encounter would take it into deeper layers, and eventually it would remain wholly inside the atmosphere. It could then make a landing as a glider, slowly shedding its speed by air resistance.

This operation (called "entry by braking ellipses") is shown in Figure 11. It is of enormous importance, as it implies that no rocket power need be used for landing on a planet with a reasonably dense atmosphere, such as Earth, Venus, and possibly Mars. A good deal more knowledge than we possess today will be needed before we can decide just what type of wing and control surfaces will be needed for the manœuvre, but the studies that have been made of the subject indicate that this kind of landing will be quite feasible. Incidentally, because of its high initial speed, the glider would have almost world-wide range.

We are now beginning to get a picture of the type of ship we will use for our first trips into space. It will be a two- or three-step rocket, with an initial weight of a few hundred tons. The take-off will be vertical, and it is quite likely that the rocket would be launched from high ground near the Equator. There are several advantages in doing this, none of them very striking but taken together they would result in an appreciable saving of fuel. A high-altitude launch would reduce losses due to air resistance, and—what is probably more important—would mean a quicker entry into the vacuum of space, where the rocket would operate more efficiently than in the atmosphere. By taking off from the Equator, the ship would already start with a horizontal speed of 1,000 m.p.h.—a modest but useful contribution to the 18,000 m.p.h. required.

At a height of about fifty miles, the rocket would be steered slowly to the east, so that the ship entered a horizontal orbit "just outside the atmosphere". (In case anyone protests against this useful phrase, rightly pointing out that there is no limit to the atmosphere, we will define it as meaning "a level where frictional resistance is so small that a body would remain safely in its orbit for as long as required". This level might be only two hundred miles up for a rocket which had to circle the Earth for not more than a few days. In the case of a permanent structure, like the "space-stations" described in Chapter 15, the minimum height might be five hundred or even a thousand miles.)

The lower steps of the rocket, discarded during its climb out of the atmosphere, would fall back to earth along a line extending eastwards from the launching site, the actual points of impact depending on the height of release and the velocities at separation. In extreme cases, the last step but one might go almost round the world before returning to Earth. This implies that the spaceship would have to be launched at the western end of a very long and narrow uninhabited region, at least if the empty steps were allowed to fall freely.

Alternatively, some kind of parachute landing might be employed, so that the lower stages of the rocket could be used again and would not be a source of danger. The best arrangement of all—and there is nothing wildly impracticable about this—would be for the individual steps to be themselves independent gliders, so that they could fly back to base under radio or piloted guidance. Certainly something of this sort would be attempted for the bottom step, which would be larger than all the others put together and thus much too valuable to lose.

Once it has become possible to establish "satellite" rockets in this fashion, astronautics should progress rapidly. Some of the scientific and even commercial uses of artificial satellites will be discussed in a later chapter: for the moment we are concerned only with their value as stepping-stones to the planets.

A rocket in a free orbit around the Earth is in an ideal situation for beginning an interplanetary voyage. In the first place it has an initial speed of 18,000 m.p.h., and therefore, as far as its fuel requirements are concerned, is more than half-way to the planets (even if it is only a few hundred miles from the Earth!). Secondly, it is in a vacuum, so that its motors can operate at maximum efficiency. And thirdly, it is in a condition of "weight-lessness".

As a result of countless stories on the subject, most people are now aware of the fact that lack of weight is a normal condition in space, though they would probably be hard-pressed to explain why this should occur. It must be made clear at once that this effect has nothing at all to do with the reduction of gravity with distance from the Earth. One could, and usually would, feel weightless in a rocket a couple of hundred miles up, where the force of gravity is practically the same as here on the Earth's surface.

The reason for this paradox may be made clear as follows. It is true enough that on the Earth gravity gives us the feeling of weight: *but it only does so when we resist it.* "G" tries to give us an acceleration downwards, but this is prevented by the pressure of the floor, the ground, or the chair in which we are sitting. Remove these supports, let gravity have its way, and we should at once feel completely weightless. We seldom notice this in practice, as on such occasions—which are brief enough anyway—our minds are usually concerned with other more urgent matters. Perhaps the only time in our normal lives when we do experience a momentary feeling of near-weightlessness is in a high-speed lift, just as it begins its descent. Parachute jumpers also know it, for a fraction of a second—before air resistance builds up and prevents unrestricted acceleration.

Now a rocket moving freely in empty space, with its motors cut off, is not resisting gravity: on the contrary, it is letting gravity take it where it will. It is sometimes said to be in free fall, but this is perhaps an unfortunate phrase because the word "fall" usually implies "downwards". The spaceship, as we have seen, could be moving upwards or horizontally while under the influence of gravity alone. So perhaps a better phrase than "free fall" is "free orbit". In this state, which would begin at the moment the rockets were turned off, the ship and its occupants would have no weight whatsoever. This condition would last even when the ship entered another gravitational field, such as when it

approached the Moon and began to fall towards it. It is therefore nonsense to speak, as many writers have done, of a steadily
increasing sense of weight as one fell towards another planet.
There *would* be an increasing gravitational force and an increasing acceleration, but the passengers in the spaceship would
feel nothing at all until the rockets were turned on again.

This matter of weightlessness is perhaps the cause of more
confusion than anything else in astronautics. One can, however,
avoid all difficulty by remembering the simple rule that, outside
an atmosphere, a spaceship and its contents are completely weightless *as long as the rockets are shut off*. It does not matter in the
least *where* the ship may be—close to a planet or in the depths
of space.

When the rockets were turned on again, the resulting acceleration caused by the thrust of the motors would produce a feeling
of weight once more—a feeling which would last only as long
as the motors were operating, and which might have any intensity
from a barely perceptible "tug" to a crushing pressure that would
prevent all movement by the passengers.

After this digression—which is fully justified by the importance
of the subject—let us return to our rocket in free orbit outside
the atmosphere. It is carrying a crew and payload, but its fuel
tanks have been practically emptied during the attainment of
circular velocity.

Now another, and similar, rocket climbs up from Earth into
the same orbit. Its payload consists simply of more fuel: one
could call it a tanker. By the use of its steering jets, it comes to
rest relative to the first ship and not far away from it.

It is this operation which causes a certain raising of eyebrows
among those who do not realise the peculiar conditions in space.
How, they ask, could two machines travelling at no less than
18,000 m.p.h. possibly make such a rendezvous?

The answer is given by the fact that this speed of 18,000 m.p.h.
is purely relative to the Earth. An observer in the satellite spaceship would consider himself at rest and the planet below would

seem to be spinning round. As the tanker rocket climbed up to meet him, it would automatically have to match his speed in order to stay in the same orbit. The whole operation would be exactly similar to that of flight refuelling in the air—the actual speeds of the aircraft being of no importance as long at they are equal. In the case of orbital refuelling, the problem of contact should be considerably easier. There is no air resistance to worry about and to affect the connecting pipe-lines: one would always have perfect visibility, and there would be plenty of time in which to make the manœuvre. If the tanker rocket aimed at a point a hundred miles away from the first vessel, and had an error in speed of a hundred miles an hour in the direction of the satellite, it would have at least an hour in which to correct this by low-powered rocket thrusts.

There are several ways in whch the actual fuel transfer could be made. The crew of the tanker could fire across a small, low-powered missile (possibly driven by gas-jets) which would tow the pipe-line behind it. It might be more convenient, on the other hand, if the two ships made direct contact, a flexible, telescopic probe on one fitting into a socket on the other—as is now done in some forms of flight refuelling.

A third possibility is that complete fuel tanks might be transferred from one ship to the other—a perfectly feasible operation in the absence of weight, and one ideally adapted for rockets built on the principles of expendable construction.

Once it had done its job, the tanker would disconnect itself, and use rocket-braking to fall back to Earth as described on page 49. After a sufficient number of round trips, the refuelling would be complete and the spaceship proper could accelerate out of its orbit and set off to the Moon or whatever its destination might be. On its eventual return to Earth, it could either enter a satellite orbit again, and the crew could be brought down to the surface by the tanker rocket, or it could make a landing itself. The advantage of the first procedure is that the spaceship would remain in its orbit in readiness for the next trip.

It will be realised that we are now a long way away from the picture of what might be called the "conventional" spaceship—a streamlined torpedo like a giant V.2, which will go direct to the Moon, land there, and come straight back without any intermediate stops. On our present knowledge, there is no likelihood of such spaceships for a very long time to come. Even atomic power, when we succeed in harnessing it for rocket propulsion, may need—may indeed demand—the use of orbital refuelling.

Another rather novel idea is beginning to emerge from this discussion. We have seen that various separate tasks are involved in the crossing of space—we have to climb up through the atmosphere, fight the Earth's gravitational field until we have got into an orbit, and then pull out of the orbit on to an interplanetary flight-path. Finally, on the return to Earth, we have to make an aerodynamic landing in a high-speed glider.

Now these various operations involve quite different—and sometimes quite opposing—characteristics in the ship that has to carry them out. For example, wings and fins are useless on a ship that has to land on the airless Moon, therefore why waste tons of precious fuel taking their dead weight all the way to the Moon and back?

The rocket which has to climb upwards from the Earth, and return as a glider, has to be stressed to withstand ten or more times the normal weight given to it by gravity, which means a strong and hence relatively heavy structure. But the rocket which spends all its operating life circling in an orbit round the Earth or some other planet, or coasting freely through airless space, could have a very much lighter construction. Indeed, by our ordinary standards it could be positively flimsy, its structual weight being only a small fraction of that needed in an Earth-based rocket of comparable size. It could thus carry greater payloads or achieve higher terminal speeds—in other words, it would be a much more efficient design.

It seems, therefore, that we are forced to the conclusion that we need not *one* type of spaceship, but at least two, depending on the job that has to be done. There is nothing new or revolutionary in this idea: it is merely sound engineering. Shipbuilders do not try to give the same vessel the characteristics both of ocean greyhound and coastal tug. If we try to do this in spaceship construction, we shall be making our already difficult task an impossible one.

In the next chapter, we will explore this idea further, and see what form the various types of spaceship may take. At this stage in the history of astronautics we cannot, of course, do more than suggest plausible designs, and new discoveries may transform our views even in the next decade. If any of the following predictions turn out to be as close to the truth as, for example, Henson's "Aerial Steam Carriage" and similar nineteenth-century forecasts of heavier-than-air flying machines, we shall be well satisfied.

6. The Spaceship

And these now earthward in their main intent
May not be found so, soon.
Minded beyond the moon
Man will enlarge his winged experiment.

EDMUND BLUNDEN—Aircraft

AN analysis of the problem suggests that we shall need three separate types of ship for the conquest of space—or, perhaps, two main types, one of which is sub-divided. They are as follows:

(A) The multistep rocket with a winged final stage, which is used to take material up to the orbit round the Earth, and then returns by atmospheric braking. We can call this a "ferry" or "tanker" ship. A very similar type of spaceship would be employed to make landings on planets with atmospheres, such as Mars and Venus. It would *not* be used for the landing on the Moon.

(B) The modified form of Type A, without wings or fins, for landing on bodies like the airless Moon by the use of rocket braking alone. It need have no streamlining of any kind. As it would employ accelerations of perhaps one-quarter of those used by Type A—since it would always be operating on fairly small planets—it need be considerably less strongly built. It would be just strong enough to stand up under its own weight on the Earth, but could not possibly lift itself in our gravitational field. It would either have to be constructed in space, or else, more probably, would be made from a Type A ship by removing unwanted structure once it had reached free orbit.

We might call this the "lunar" type spaceship. It would

be used to land on Mercury, the Moon, and the satellites of the giant planets. Should the Martian atmosphere prove too thin to permit of an aerodynamic landing, we may also have to use it there.

(C) Finally, we have the "deep space" type of ship which never lands on *any* planet. Its function is to travel from an orbit round the Earth to an orbit round the Moon, Mars, or any other world, carrying personnel, fuel and stores. At its destination it would be met by a Type A or B ship which would make the transfer to the planet beneath. The ferry rocket would, of course, have to be carried to the planet by earlier Type C ships, or it might have travelled there previously under its own power.

The "deep space" ship we are trying to visualise would be a very curious-looking contraption. Probably if we saw a photograph of one we would not realise we were looking at a spaceship at all, so alien might it be to our present-day ideas. Like the "lunar" type of ship, it would have no vestige of streamlining and could be of whatever shape engineering considerations indicated as best. But, unlike any other kind of spaceship, it would never have to withstand more than very small forces and accelerations. Floating in its orbit around the Earth, it could build up speed to escape velocity and spiral outwards into space in as leisurely a fashion as it pleased—taking perhaps hours or even days over the process, instead of minutes as in the "conventional" type of spaceship.

We have never, on this Earth, had the opportunity of building structures which need not even be strong enough to stand up under their own weight. The only forces the "deep space" type of ship would ever have to experience would be those due to its gentle acceleration and any turning manœuvres that proved necessary. Probably these would not amount to more than a tenth of the weight which earth-gravity would have given the structure, and the ship might consist of a small, compact crew chamber and a series of very large and apparently fragile fuel tanks, the whole held together by an open lattice-work of spidery struts and spars. It would have about as much structural strength as a Chinese lantern, and perhaps the analogy is not a bad one as the tanks could, at least for some fuels, be little more than stiffened paper bags!

The rocket motors themselves would be very small and relatively weak, designed to produce modest thrusts over considerable periods. This again would produce a very substantial saving of weight.

An attempt has been made, in Plate II, to show the various types of spaceship we have envisaged. In the foreground a winged, Type A rocket is refuelling a Type B ship which is destined to land on the Moon. Both ships, it should be realised, are the upper stages of considerably larger rockets, the lower steps of which never escaped from the Earth.

As has been explained on page 52, the ships and any objects moving with them are under conditions of zero gravity. Neither the direction towards the Earth nor any other direction is "down". The members of the crew who may leave the ship in their spacesuits to assist with the fuel transfer will be weightless and can float where they will under the impetus of their reaction-pistols. As an additional safety factor, they would probably use lines attaching themselves to the vessels, so that there could be no danger of drifting away into space.

The "lunar" or Type B ship is built to withstand accelerations of one gravity, or six times that produced by the Moon's field. The landing gear is already in place since there is, of course, no point in making it retractable. When the ship is on the Moon (see Plates III and IV) the "undercarriage" would play the rôle of a launching rack, holding the rocket in the required position for take-off. It could, therefore, be left behind, since it would not be needed on the return to the orbit around the Earth, and so might be made detachable. However, this would be bad economics, because it would be cheaper to bring it back than carry up a new set of landing gear from the Earth when the ship had returned and was preparing for its next voyage. Nevertheless it would give the spaceship a large and valuable safety factor, since in an emergency the undercarriage could be left on the Moon. (Where it would no doubt provide very useful material for the colonists at a later date!)

An "outrigger" near the forepart of the ship carries the equipment necessary for visual and radar observation of the approach to the Moon. This apparatus has to be kept clear of the rocket jet, which would not only produce such an intense glare that an observer would be blinded, but would also absorb the radar impulses rather readily in its column of ionised gases.

In the distance is the Type C ship with a refuelling tanker which has done its job and has now uncoupled preparatory to returning to Earth. The pressurised crew chamber and fuel tanks are in one sphere, the motors in the other.

Although many alternative layouts are possible, this "dumb-bell" arrangement has certain advantages. From the engineering point of view, the sphere is the shape which lends itself most readily to pressurisation, besides having other structural virtues such as least surface area for greatest enclosed volume. The two spheres are linked by a connecting cylinder which carries the controls and acts as an access tunnel.

One advantage of this arrangement, which might be decisive, is that it would seem ideal for atomic propulsion. We will discuss this subject on page 63, but for the present it is only necessary to say that any type of "atomic rocket" which may eventually be constructed is likely to produce large quantities of dangerous radiation. Shielding of the crew will therefore be essential, and this will be provided to some extent by the fuel itself. In addition, by having the crew chamber as far away from the motors as possible, the additional safety of distance is provided at relatively little cost.

It will, of course, be realized that ships of this type would have to be constructed in space from components ferried up from

Earth and assembled in free orbit. As exactly similar problems will be considered when we discuss the building of "space-stations", we will not go into them here, except to remark that constructional difficulties should be greatly eased under conditions where nothing has any weight and so will stay where it is placed, with no visible means of support!

It will also be understood that this type of spaceship is not likely to be built until some years after the Types A and B, which will carry out the first exploration of the planets and will prepare the way for the later and more efficient ships.

In the absence of any revolutionary technical development (and perhaps it would be wise to remind ourselves that this is a somewhat improbable assumption!) it seems, therefore, that the conquest of space will take place in the following stages:

(1) Unmanned, instrument-carrying missiles will enter stable orbits round the Earth, and will travel to the Moon and planets.

(2) Manned, single-step rockets will ascend to heights of several hundred miles, landing by wings or parachutes.

(3) Multistage, manned rockets will enter circular orbits just outside the atmosphere and, after a number of revolutions, will return by rocket-braking and air resistance.

(4) Experiments will be made to refuel these ships in free orbit, so that they can break away from the Earth, make a reconnaissance of the Moon, and return to the Earth orbit.

(5) The type of ship designed for a lunar landing will be flown up from Earth or assembled in free orbit, and after refuelling will descend on the Moon. The ship may then return direct to an orbit around the Earth, or it may make a rendezvous, *in an orbit round the Moon,* with tankers sent from Earth.

(6) While the exploration of the Moon is proceeding by the use of such ships and techniques, attempts will be made

to refuel rockets for the journeys to Mars and Venus. Although the power requirements are not much greater than for the lunar voyage, the duration of flight (145 days to Venus, 240-250 days to Mars) would demand the use of considerably larger vessels. Whether any attempt would be made to land, on these first trips, would depend on several factors which cannot be foreseen today—in particular, upon the new knowledge gained by reconnaissance rockets in the meantime. Obviously, it would be very desirable to make a landing, in view of the expense and duration of these voyages, but this would increase the initial size of the ships very greatly. The first expeditions might have to be content with an orbital survey before returning to Earth, filling in the gaps left by the earlier reconnaissance missiles.

(7) Finally, landings will be made on Mars and Venus, perhaps by specially designed and quite small vessels which will accompany the main ship. These little "tenders" or "ferries", when they had carried out their task of taking selected members of the spaceship's crew down to the planet, and bringing them back to the orbiting mother-ship, would be left behind, still in free orbit, for the use of future expeditions.

At this stage, the first era of interplanetary flight would be ended. Thereafter the problem would be one of improving the efficiency of spaceships, building up bases on the Moon, Mars and Venus, accumulating stores of fuel at the most useful places (in free orbit as well as on the planets), and preparing for the longer and technically more difficult journeys to the giant outer worlds and their satellites.

It will be realised that the various advances outlined above will overlap to a considerable extent. Moreover, if it becomes possible to employ atomic energy for propulsion, some of the intermediate stages may be rendered unnecessary and the whole progress of space-travel accelerated. Many authorities believe that such a development is highly probable, but it cannot yet be regarded as certain and it would be unwise to base all one's hopes upon it.

At the present time, atomic energy has been released only in the forms of radiation and heat. It can be produced at all power levels from that needed to boil a cup of water in five minutes to that required to smash a city in a few seconds. There is no way, however, in which it can be used *directly* to provide a controllable thrust of the kind needed for propulsion. To do this, we must employ some roundabout method of converting energy of heat into energy of motion.

The conventional rocket, of course, does exactly this. During combustion, the chemical energy of the fuel is liberated, and so heats up the various gases produced in the reaction. During their expansion and progress down the nozzle, these gases, to put it picturesquely, exchange heat for speed. They become cooler, but acquire a high velocity and so generate the thrust which propels the rocket.

It is clear, therefore, that if we could heat up a gas and let it expand in this fashion, we would have a type of rocket. The heat need not come from a chemical reaction: its origin is of no importance, and a nuclear reaction would do just as well. This means that we could use *any* gas for our propulsive jet—it need not be one that takes part in some combustion process. Our freedom of choice is thus greatly increased: we could even use, if we wanted to, a gas such as helium, which is completely inert and undergoes no chemical reaction at all.

An "atomic rocket", therefore, might consist of these components: (a) a tank of propellant—probably hydrogen, (b) a nuclear reactor or pile which can operate at very high temperatures, (c) a heat interchanger for transferring the thermal energy from the pile to the propellant gas and (d) the rocket nozzle. Note that the place of the combustion chamber in the conventional rocket has been replaced by the pile and heat interchanger.

This arrangement sounds very simple in theory, and—also in theory!—it has some most attractive advantages. In the first place, all the energy we need can be carried in quite a small atomic pile, for nuclear reactions liberate something like a *million* times the energy of chemical reactions. It is now possible to build piles (employing enriched uranium instead of the element in its normal form) which are only a few feet on a side, and capable of operating at any power level. The great problem is getting the heat out of the pile quickly enough to prevent it melting—and then to transfer this heat to the propellant.

That is the purely engineering problem, and it is an exceedingly difficult one. But there is also the question—unavoidable when dealing with atomic energy—of dangerous radiations. The nuclear reactor itself would become "hot", to use the expressive slang which has now passed into the language of physics, as soon as it started to operate, and it would remain dangerously radioactive even when closed down again. This means that it would have to be heavily shielded, at least on the side towards the crew. The studies that have been made of the subject indicate that, if the spaceship was a long, narrow structure with the motor at one end and the crew at the other, the weight of shielding would not be prohibitive.

The problem does not, however, stop here. Under certain circumstances the rocket exhaust would also be dangerously radioactive, and this is an even more serious matter, as the rapidly moving gases could contaminate a very large area. This fact may well make it unsafe for nuclear-powered rockets to take off from the Earth's surface, but would not rule out their use in space. Once again, therefore, we return to the idea of orbital spaceships. Chemical rockets could be used to climb up from Earth and reach orbital velocity, but from there onwards atomic motors would be used. The actual landing on another planet might employ chemical rockets once more, to prevent local contamination by the blast.

Using the atomic rockets in this manner might be a matter of necessity for yet another reason. Although this is still largely speculation, it seems likely that the atomic drive may be better adapted to produce small thrusts for long periods of time than very great thrusts for short periods. We have to use the latter to get our spaceships away from the Earth, and thus we may be forced to employ chemical rockets at this stage. But once in space, when the gentlest of drives will eventually produce any required velocity, atomic propulsion may come into its own.

The picture we now have of the various vehicles needed for interplanetary travel is a good deal more complex than has been suggested, for example, in the many works of fiction the subject has inspired. Yet, when one considers the matter, it would hardly be otherwise. Space-flight involves many different problems, varying greatly with different missions. One might therefore expect to find almost as many diverse types of spaceships as there have been types of aircraft—and one day, no doubt, there will be an equally bewildering variety of designs.

Some features, however, all spaceships will have to possess in common. They will have to provide a comfortable environment for their occupants, supplying them with air and maintaining them at the correct temperature, irrespective of surrounding conditions. Adequate food and water supplies will have to be carried in as compact a form as possible. Every ship, once it has been launched on its journey, will be a tiny self-contained world relying entirely on its own resources: those aboard it can expect no help from outside if anything has been forgotten or if there is a failure in the ship's mechanisms. Complete reliability and self-sufficiency will be the targets for which the spaceship designer must strive, once the minimum requirements of fuel and payload have been met.

Let us first consider the problem of the air supply. Here at sea level on the Earth, we are under an atmospheric pressure of about fifteen pounds per square inch, or almost a ton for every square foot of our bodies. We are not normally aware of this pressure, because it is equalised inside and out. Given time to adjust itself, the body can function over a considerable pressure range—down to a third of the normal atmospheric value, and up to four or five times this. The actual limits depend on the length of time the abnormal conditions last: clearly in a spaceship which might be travelling for weeks or months the pressure must be kept at a value comfortable and safe for the crew. There is no need, however, for it to be as high as the standard sea-level value of fifteen pounds per square inch, and indeed there are sound reasons why it should be as low as practicable.

The cabin of the spaceship has no equalising pressure outside it, being in a perfect vacuum, and so it must be strong enough to withstand the full internal pressure. To build a large container which will not burst when there is a force of one ton acting outwards on every square foot of its surface is not easy, particularly when weight is at a premium. The use of a lower pressure would simplify construction, and would also reduce the small air loss through the leaks which are inevitable in any pressurised system.

Fortunately, there is no difficulty in doing this. Our normal air is only one-fifth oxygen—the remaining four-fifths of nitrogen is simply "ballast" and plays no part in respiration. Thus the oxygen in the atmosphere contributes only three pounds to the total fifteen pounds of pressure—and if we used in the spaceship an atmosphere of *pure* oxygen at three pounds pressure our lungs would receive just as much of the gas as under normal conditions.

It is not yet certain if it is quite safe to live indefinitely in a pure oxygen atmosphere at three pounds per square inch—a fifth of normal pressure—but it is undoubtedly safe for prolonged periods. This fact is of great importance not only in the design of spaceship cabins but also of "space-suits", where the same problems have to be met, together with the additional requirement of flexibility.

Having decided to employ a pure oxygen atmosphere, we must provide the spaceship with some means of removing the carbon dioxide produced by respiration. Several chemicals are known which can perform this feat (e.g. sodium hydroxide and sodium peroxide), and the latter will not only remove the carbon dioxide but will replace it with fresh oxygen. In addition, supplies of pure oxygen can be carried in the liquid state in suitable storage flasks.

The amount of oxygen needed by a man under normal conditions is surprisingly small—just over three pounds a day if he is engaged in continuous moderate exertion. When resting or sleeping, the consumption is reduced to a third of this value, and as there would not be much physical activity inside the ship, an allowance of two pounds per man per day would appear to be ample.

It would also be necessary to remove excess water vapour from the air. This can be done by chemical means, but a simple and effective method is to pass the air through a chilled pipe and condense the water out of it.

On very long voyages, the chemicals needed for oxygen replenishment would weigh a considerable amount, and it has been seriously suggested that we might employ Nature's method of purifying the atmosphere—in other words suitable green plants should be carried in the spaceship! As is well known, plants absorb carbon dioxide (in the presence of sunlight) and, after converting it to starch, liberate oxygen. Although this idea is an attractive one at first sight, it loses some of its charm when one considers the additional complications and the weight of chemicals that would have to be taken along to feed the plants.

This scheme, or some variant of it, may, however, be used in certain cases, for biological processes can often perform feats beyond the power of the chemist. No simple, direct way is known of converting carbon dioxide back into oxygen, yet this is a task performed by every blade of grass in the world. On space-stations, and in the bases we shall set up on the planets, the atmosphere may well be kept pure by the use of plants specially bred for this purpose.

At this point we might mention one curious result of the absence of gravity which has an unexpected effect on the air-conditioning problem in spaceships. The gases which we exhale, being considerably warmer and hence lighter than the surrounding air, normally rise upwards so that, even when we are sleeping or sitting still, the air around our nostrils is continually replenished. This effect is seen most clearly in the case of a candle flame, which has a steady current of fresh air flowing into it from below. Now this form of circulation, since it depends on differences in weight, cannot occur aboard a spaceship in its normal, free-orbit condition. It has, in fact, been shown experimentally (by filming candles in a freely falling chamber) that flames cannot burn in the absence of gravity: they quickly "suffocate" in their accumulated combustion products.

This implies that an efficient system of forced ventilation must be installed in a spaceship to sweep away the waste gases as soon as they are formed. It is also a warning that one must take nothing for granted in space, and a reminder that gravity may be an important factor even in processes that seem unconnected with it.

Next in priority to the air supply is the regulation of temperature. The "temperature of space" is a subject about which there is much confusion, the general idea being that it is extremely cold outside the atmosphere. In fact, the reverse is nearer the truth.

Consider the case of a solid body floating in space at the Earth's distance from the Sun. One side will be in shadow, the other in full sunlight. This side will become extremely hot—at least if it is darkened and so readily absorbs heat waves. In the extreme case of a completely black body, the temperature at the parts directly facing the Sun would be somewhat above that of boiling water. On the other hand, if the surface was white or silvered, it would reflect most of the heat and so be quite cool.

The dark (or night) side of the body would be cold in any case, since it is continually losing heat by radiation and has none coming in. However, if the body were a good conductor, the temperature extremes would be equalised to some extent. And if, as in practice would probably be the case, it were rotating even at quite a slow rate, the temperature over its whole surface would be practically uniform.

It must also be remembered that a considerable amount of heat would be generated inside the spaceship by the bodies of the passengers. A double-hulled ship in space would behave like a very efficient thermos flask, and it would often be more important to *lose* heat than to conserve it.

On a journey from Earth to Venus, the amount of heat received from the Sun would be twice as great at the end of the voyage as at the beginning. On the trip to Mars, the reverse would be the case. Moreover, when the ship was in the shadow of a planet—which would be for a considerable fraction of the time if it were circling in a close orbit—there would be no solar radiation at all. This last condition would also apply at night on an airless world such as the Moon. It appears, therefore, that the ship should have some simple but efficient means of regulating its heat-loss, and this might be done by the use of folding shutters which would expose blackened or silvered areas of the hull according to circumstances. An internal source of warmth would be needed during prolonged periods of darkness, and this could best be obtained by the combustion of fuel in a small heater.

In general, it can be said that temperature control does not present very great problems on the voyages which we will be undertaking during the first decades of space-travel. On journeys very near the Sun, or far beyond Mars, the situation will be different. However, by the time such feats as these are seriously contemplated, we should have plenty of power from nuclear sources to use either for heating or refrigeration.

We shall return later to some of the other problems of life in space, such as navigation, steering, communication, and so on. Before closing this chapter, however, it is worth noting that aboard a spaceship, since there would be no "up" or "down", or indeed any preferred direction except during the short periods of acceleration, the designers of the crew cabin would have a freedom which terrestrial architects might well envy. Walls and floors would be interchangeable, and the whole volume of enclosed space available for any purpose. Only spaceships which had to take off from Earth, or which would be used as headquarters on some other planet (see Plate IV), need have cabins designed with a definite "up and down" direction.

This is assuming, of course, that space-travellers can grow accustomed to the lack of gravity and that it has no unfortunate physical effects. This important point is discussed in Chapter 9, so we need only remark here that all the medical evidence suggests that "weightlessness" will not be a dangerous condition—and even if it is found to be, there is a simple way of overcoming the difficulty.

7. The Journey to the Moon

But the principal failing occurred in the sailing
And the Bellman, perplexed and distressed,
Said he *had* hoped, at least, when the wind blew due East
That the ship would *not* travel due West!

LEWIS CARROLL—*The Hunting of the Snark*

WE will now consider in some de-
tail the first of all interplanetary voyages—the journey to the
Moon. The human race is remarkably fortunate in having so
near at hand a full-sized world with which to experiment: before
we aim at the planets, we will have had a chance of perfecting
our astronautical techniques on our own satellite.

The spaceship starts, we will assume, from an orbit close to
the Earth, after having been refuelled as shown in Plate II. Since
it is travelling round the Earth in a period of about ninety
minutes, it is completely changing its direction of motion every
three-quarters of an hour. If it is travelling westwards at its
orbital speed of 18,000 m.p.h. at one moment, it is moving eastwards at the same speed forty-five minutes later. This, of course,
gives the pilot a very wide range of control, though the timing
for "take-off" must be exact.

For the ship to reach the Moon's orbit, it must leave the vicinity
of the Earth at a little less than full escape velocity—24,900
m.p.h. instead of 25,000 m.p.h. The difference may appear to be
trivial, and so it is as far as fuel requirements are concerned.
However, the time of flight depends very critically on the initial
speed—an increase of only one per cent. in the minimum speed
more than halving the duration of the voyage!

Starting at 24,900 m.p.h. from near the Earth, the ship would
reach the Moon's orbit 116 hours later: starting at 27,000 m.p.h.
it would cover this distance in nineteen hours. The reason for

71

these surprisingly drastic reductions may be understood by looking again at Figure 9. The minimum speed of 24,900 m.p.h. would barely get the ship to the Moon, high on the horizontal slope of the "pit": for the last scores of thousands of miles of its journey it would be creeping along very slowly. If, however, it started with anything in excess of the minimum speed, it would still have a considerable velocity as it approached the Moon and would no longer be covering great distances at very low speeds.

We will assume that the spaceship draws away from its circular orbit at an acceleration of one gravity. After just over five minutes, it would have built up speed from 18,000 m.p.h. to the 24,900 m.p.h. needed to get it to the Moon. During the closing stages the motors would probably be throttled back to give a lower acceleration, in order to make it easier to make the final velocity adjustments.

The whole manœuvre would be carried out according to a prearranged programme, in order to put the ship into an orbit that had been calculated long in advance. It is very probable that the entire operation would be automatic and monitored by radar stations and observatories on Earth, which could give a continuous check of the ship's position and velocity, and would radio correcting signals to the automatic pilot.

The five minutes of acceleration over, the ship would then have five days of free coasting ahead of it. This would give ample time for any further checks on its course, and such small steering corrections as proved necessary could be made at the appropriate times.

On the fourth day, the Moon would be so near that its attraction would begin to produce an appreciable effect on the ship's course. (Once again it might be mentioned that those in the ship would be unaware of this, except from the evidence of their instruments.) About 24,000 miles from the Moon, the gravitational fields of Earth and Moon balance. Up to this point, the ship would be losing speed all the time (even though it would still be rising away from Earth), but afterwards it would commence to gain speed—towards the Moon. If allowed to fall unhindered, it would crash into the Moon at about 5,200 m.p.h.

The problem of landing on the Moon is one which, despite its obvious importance, has not received as much attention as it should. Clearly it can only be done by rocket power, for there is no question of air-braking or the use of parachutes. It would appear that the most economical method of approach is to fall directly towards the Moon, using the rockets at the last possible minute to check speed in one burst. The landing would be, in fact, exactly like a take-off in reverse.

It will therefore be necessary, while still some distance from the Moon, to "reorientate" the ship so that its motors are directed towards the Moon, and this raises a point which is worth a little discussion.

The direction along which the axis of a coasting spaceship points (its "attitude", as it is called) is completely independent of its direction of motion—which is itself a quite arbitrary quantity, depending entirely on the viewpoint of the observer who measures it. Unless it were deliberately stabilised in some way, a spaceship would normally be rotating and perhaps turning end over end as it travelled along. We have seen that this would be advisable to equalise temperature, though it would make observation and radio communication difficult. It would be easy to "kill" these rotations, if required, by the brief use of small rocket jets aimed in the appropriate directions, and such jets could also be used to give the ship any desired orientation.

This, however, is not the only way in which the attitude of a spaceship might be controlled and altered. The other method is by the use of a gyroscope or large flywheel, and both systems are illustrated in Figure 12.

The upper series of sketches show how a ship spinning in space in a "clockwise" direction could have its rotation stopped by a rocket fired against the direction of spin. This method would be

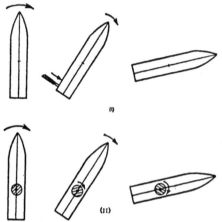

(i)

(ii)

Figure 12. Checking the Spin of a
Spaceship (I) by Rocket, (II) by Flywheel

useful to get rid of
unwanted rotation, but
might be inconvenient
if one only desired to
turn a *non-spinning* ship
through a definite angle
and thus make it point
in a new direction. To
do this by rocket power
alone would be a tricky
operation, since the
rockets would have to
be used twice—once to
start the ship turning,
then again to check its
spin when it had rotated through the required angle.

The second method of control is shown in the lower row of
drawings. Suppose there was a flywheel at the ship's centre of
gravity, and imagine that at first it is fixed relative to the slowly
turning ship. If now a motor starts to spin the flywheel in the
direction shown, as the flywheel builds up speed the ship will
gradually cease its rotation—until eventually it has come to rest
and all the spin has been transferred to the flywheel. Since the
flywheel will be very much smaller and lighter than the ship,
it will obviously have to rotate at a very high speed to be effective,
which means that it could not deal with large rates of spin. Note,
also, that it would have to be kept running: if it slowed down, it
would transfer its spin back to the spaceship!

The flywheel method would therefore be best suited for
changing the orientation of a non-spinning ship. For this purpose,
it would be started up, thus making the ship turn very slowly,
and then stopped again as the ship came round to the required
position.

It would take many minutes to "reorientate" a large spaceship by the use of flywheels (or, what comes to the same thing,

gyroscopes) of any reasonable size. However, there would be hours or even days available in which to make these manœuvres, so this is no great handicap. In the above illustrations we have considered motion in only one plane, and of course there are three possible directions of spin: but this does not affect the argument. It means that we would have to use either three fly-wheels at right angles, or—more probably—one flywheel which could be set spinning in any plane.

We might mention here that the movement of the crew inside a spaceship would produce changes of spin and orientation, though very small ones. If it proved necessary, these could be taken care of by automatic, gyro-controlled equipment.

To return to our ship falling towards the Moon. We will assume that, by one or both of the above methods, we have turned it in space so that the motors are aimed at the Moon. Using the rockets to produce a deceleration of one gravity, a speed of fall of 5,000 m.p.h. could be checked in four minutes, during which time the ship would cover a distance of a hundred and sixty miles. It would certainly seem dangerous—not to mention hard on the nerves of the crew!—to wait until the Moon was so close before doing any braking. Yet it would be quite safe, especially if there were even a small reserve of power for in-creased deceleration. (At 2 g, which is still a modest figure, the ship could be brought completely to rest in less than two minutes after falling some eighty miles.)

The landing, like the take-off, would almost certainly be auto-matic. A radar altimeter would give the exact distance to the Moon as well as the rate of descent, and this information would be passed to an electronic computer which would control the motors. The ship would then descend according to a prear-ranged programme that would bring it to rest a few feet above the Moon.

It is rather natural to imagine that a rocket descending tail first in this manner would be a highly unstable affair and liable to "topple". But it must be remembered that, as far as the automatic

controls are concerned, there is no fundamental difference between a 1-g vertical landing and 1-g vertical take-off. If the
gyroscopes and steering devices can deal with the one case—as
they already do in the V.2 rocket—they can deal just as well with
the other.

What sort of "undercarriage" our lunar spaceship would need
is an interesting problem which cannot, perhaps, be completely
solved until we know a good deal more about the surface of the
Moon. By the time a landing is attempted we will certainly have
large-scale photographs of the lunar plains and will have located
numerous relatively flat areas which would provide suitable
places for a touch-down.

Helicopter undercarriages have been made which are able to
absorb rates of descent of forty feet a second, and this would
correspond to a free fall on the Moon from a height of a hundred
and sixty feet. Thus if the spaceship could be brought to rest
twenty feet from the lunar surface there would be no difficulty
in providing landing gear which would stand the impact without
being too bulky and heavy. It might resemble that shown in
Plate III: even the low lunar gravity would demand something
more substantial than the fragile fins or wings which, in so
many magazine illustrations, seem to be the only means of support
with which spaceships are provided!

An impression of the ship on the Moon, with a second vessel
descending on its braking rockets a few miles away, is given in
Plate IV. One space-suited member of the crew is making a film
of the landing. To the left of the ship is a radio mast (a collapsi-
ble rubber structure blown up by air pressure) which enables the
explorers to keep in contact with the base as they investigate the
local landscape.

We will reserve discussing the problems of living on the Moon
until Chapter 11, and go straight on to consider the return
journey. When it is ready for departure the ship will, of course,
have left behind all expended stores and any equipment not
needed for the return. The take-off may be made whenever de-

sired: there is no advantage in choosing one time rather than another. (In theory, it is slightly easier to make the return when the Earth and Sun are pulling in the same direction, but the saving in fuel is completely negligible.)

Two possibilities now arise. The ship may build up speed at once to lunar escape velocity (5,200 m.p.h.), which, it will be remembered, could be reached in four minutes under only one gravity. It would then travel back to Earth along an orbit similar to that which it traversed on the outward voyage, and taking the same time—five days. This would be the simplest procedure, if the ship carried sufficient fuel. On the other hand, it is more than likely that the first spaceships to land on the Moon would do so with very slim fuel reserves, and might only be able to reach orbital velocity. In the case of the Moon, this is 3,700 m.p.h. (thus a V.2 rocket could quite easily become a lunar sub-satellite!). The ship might then be refuelled either by a tanker from Earth which had been waiting for it, or it might make a rendezvous with orbiting fuel tanks which it had itself left in space before descending to the Moon.

There is a good deal to be said for this latter procedure, since obviously it is a complete waste of effort to carry the fuel needed for the full return journey down to the Moon and then lift it up again. It remains to be seen, however, whether this idea will prove to be practicable when allowance is made for all the complications it involves, and the restriction it puts on the return orbit.

During its five-day fall back to Earth, the ship will build up speed (at least, once it has passed the "neutral point" where the two gravity fields balance) and if it were not checked would reach the Earth again at the velocity with which it began its journey—24,900 m.p.h. As the type of ship we have been discussing would return to a circular orbit, and not make an actual landing on the Earth, it would have to shed just under 7,000 m.p.h. of its velocity to reach the speed of 18,000 m.p.h. needed for a close orbit. It could do this by means of air-resistance braking—as described on page 49—and the use of a small amount of rocket power to adjust the final orbit as necessary. At the end of these manœuvres, which would occupy only a few hours, it would be back in a stable, circular orbit waiting to be refuelled and serviced, and the crew could be taken down to Earth by one of the winged "ferry" rockets.

As has been stated before, the whole problem of the lunar journey—as of any interplanetary voyage—would be enormously simplified if it were not necessary to carry fuel for the return journey. Even if we avoid the need to build a single impossibly large ship by arranging a rendezvous with a tanker off the Moon, it would still be necessary to carry many separate loads all the way from the Earth, with the great expenditure of fuel which that implies. The entire economics of space-flight would alter drastically if it became possible to *refuel on the Moon*. This may seem a fantastic assumption, and clearly it would not be practical until a lunar colony, with considerable plant and equipment, had been established. But if space-flight is to be of any value it will lead to exactly this sort of thing. One of the first aims of such a colony would be, in fact, to search for material which might provide rocket fuel.

The enormous potential importance of the lunar base will be discussed in Chapter 11, but it is worth pointing out here that there is every reason to suppose that all the elements found on Earth will also exist on the Moon, though no doubt in different compounds and distributions. So the materials needed for any desired fuel would be available, if the problems of mining, purification and so on can be solved. In particular, it is highly probable that water can be found on the Moon—not, of course, in the liquid state, but either frozen or chemically combined. Its extraction would be a fairly straightforward process, and it could be used to provide the oxygen and hydrogen needed for a chemically propelled rocket. What is an even more interesting possibility, it might be used *directly*, without any processing, as the propellant for an atomic rocket. From the viewpoint of sheer performance, pure water would not be as good a "working fluid" as hydrogen, but availability and ease of storage might well outweigh this disadvantage, at least as far as spaceships returning from the Moon were concerned.

In these circumstances, a not-very-efficient atomic rocket could be a much more attractive proposition than a chemical rocket burning fuels which were difficult and costly to prepare—and the Moon would become, in an almost literal sense, our stepping-stone to the planets.

8. Navigation and Communication in Space

"Other maps are such shapes, with their islands and capes!
But we've got our brave Captain to thank"
(So the crew would protest) "that he's brought us the best—
A perfect and absolute blank!"

LEWIS CARROLL—*The Hunting of the Snark*

WE have already discussed, in Chapters 5 and 6, the types of orbit which spaceships will have to follow in order to reach the planets. The problem, it will be recalled, is basically one of attaining the correct velocity at the correct time, and then waiting until the period of free coasting brings the ship to its destination. In this chapter, we will investigate some of the subsidiary but nevertheless important questions which this involves.

One of the most interesting of these is the subject of navigation —which includes determination of position and velocity, as well as the actions required to make any necessary corrections to the course. Although this is a subject of very great complexity—if one goes into it in any detail—some of the answers can be expressed fairly simply.

In one respect, a spaceship is at a great advantage over most terrestrial vehicles. It is always in conditions of perfect visibility —the entire heavenful of stars is available for observation. The stars themselves provide a kind of fixed framework against which to relate one's observations, but they are so far away that, by themselves, they would do no more than this. To locate the actual position of the spaceship it is necessary to take observations of the Sun and planets.

The navigator of a spaceship, like that of any other vessel,

would have an almanac
or "ephemeris" containing
tables which gave the positions of the planets at any
time, and Figure 13 shows
how he could use this to
locate his ship. (For simplicity, we will assume that
ship and planets lie in the
same plane. This would not
be very far from the truth,
and there is no difficulty in
allowing for the error in
practice.) Let us suppose
that Venus and the Earth
are the two most conven-
ient planets to observe.

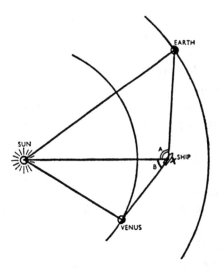

Figure 13. Position-finding in Space

By means of a sextant, or whatever its equivalent device may
be in astronautics, the navigator first measures the angle A between Sun and Earth. He knows, from the almanac, the position
of the Earth, and hence the line Sun-Earth is fixed. Next he
measures the angle B between Venus and the Sun—and since the
line Sun-Venus is also known, simple geometry fixes the spaceship's position at X.

Since there would seldom be less than three, and would
often be five, bright planets available for observation this is
clearly a very useful method, as well as an extremely simple
one.

An alternative though less accurate method of position-finding
would be to measure the *apparent sizes* of the Sun and planets.
Since their diameters are known, this would fix the *distance* of
the ship from each of them, and so locate it in space. This procedure would be particularly valuable when one was approaching
a planet, and its disc was becoming fairly large. Simple sextant
measurement would give one's distance with accuracy, until the

ship was so close that a low-powered radar set could take over
the task of range measurement.

His position is, clearly, only one of the things a space-navigator
would want to know. Equally important is the velocity of the
ship. This could be found by another measurement of position
at a later time, but any way of discovering velocity at once would
be very valuable. There seems no simple astronomical method
of doing this, unless use can be made of the Doppler effect—
the alteration of the frequency of light caused when one is moving
towards a source or away from it. This effect is extremely small
at the sort of speeds interplanetary spaceships would attain, and
it would be very difficult to measure it accurately.

A much more practical method would involve the use of
radio stations on the planets. To take a simple case, suppose
the spaceship were near the Earth and a terrestrial station were
sending out a series of "pips" at regular intervals—say at the
rate of a thousand a second. Then if the ship were travelling
towards Earth, it would receive more than a thousand pips a
second: if away from it, less. Modern electronic equipment can
make such measurements with great accuracy and one day, no
doubt, there will be radio beacons on the planets and possibly
in space for the assistance of "astrogators".

Once a spaceship has ascertained its position and its velocity,
or has made two accurate "fixes" at a sufficient interval apart,
its future position can, in theory, be completely determined from
the laws of celestial mechanics. Until the motors are turned on
again, it will travel through space as a captive of the Sun, no
more able to escape from its orbit than can the planets or comets.
The actual calculation of the orbit would, however, be a complicated task and might require the use of electronic computers
altogether too large to be carried in a spaceship. Possibly this
duty would be relegated to machines on the Earth, which would
transmit their results by radio to the spaceship within a few
minutes of having received its observation.

At fairly long intervals during the course of a journey it would become apparent that the ship's actual position was beginning to deviate from the orbit it should be following, owing to the inevitable errors in the initial period of powered flight. It might then become necessary to alter the ship's course in speed, direction, or both, to correct for this.

We have already pointed out, on page 73, that the actual orientation or attitude of a spaceship has nothing to do with its direction of motion: it could be spinning or pointing in space in any manner without affecting its velocity. To alter this, the rockets have to be used in the manner shown in Figure 14.

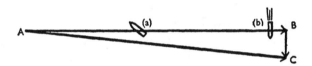

Figure 14. Altering the Velocity of a Spaceship

Suppose that the velocity of the ship is represented, in magnitude and direction, by the line AB. (Such a line is called a vector.) At the moment we are concerned with, the ship's attitude is that shown at (a). Now let us suppose that calculation has shown that the velocity *should* be along the line AC. Elementary dynamics tells us that to give the ship this required velocity, we must impart to it a speed in the direction BC, and of magnitude proportional to BC. This would have to be done by turning the spaceship into the position shown at (b), and firing the rockets for the calculated period of time.

It should be emphasised that, under normal conditions, the change in course required would seldom be more than a degree and so the amount of power needed for such corrections would be extremely small.

At this point someone may very well ask: "Exactly what is meant by the *velocity* of a spaceship? From what reference point does one measure it?" This is a very important question, for a little thought will show that it is meaningless to speak simply of a spaceship's velocity without adding a little more information.

For consider a ship between Earth and Mars. Observers on Mars, after a series of careful measurements, might decide that the ship was approaching them at 12,570 miles an hour. Equally accurate observations from Earth might show that it was receding at 8,490 miles an hour. Measurements by observers on the other planets would give different results again, since no two points in the Solar System are at rest with respect to each other. All the measurements would be equally "correct": which is the spaceship to use?

The answer is purely a matter of practical convenience. A spaceship leaving the Earth is primarily interested, at least during the first few hours of flight, in its speed with respect to the Earth and would use that body as its reference point. After a few days, when it was well on its voyage and the Earth's gravity was already very feeble, the Sun would be the obvious "landmark" to use—all the more so as the ship's path would now be controlled almost entirely by the Sun's attraction. When the planet of destination approached, there would be another change of reference system, for the navigator is now chiefly concerned with the manœuvres which will bring him in to land.

A good terrestrial analogy is given by the case of an airplane flying from one aircraft-carrier to another. During the moments of landing and take-off, the pilot is chiefly interested in his speed over the flight-deck. When he is airborne and above the sea, what matters then, for navigation purposes, is his ground speed.

In all our discussions so far, we have assumed that the orbits of the planets, and that of the spaceship, lie in the same plane. This is only a "first approximation", as the mathematicians would call it. The orbits of Mars and Venus, the two planets in which we are most interested, are actually inclined to the Earth's at angles of 1.9 and 3.4 degrees respectively. These values may seem very small, but because of the great distances involved they would, if they were ignored, cause a spaceship to be "above" or "below" the planets, at the end of its voyage, by as much as 5,000,000 miles. To correct for this it would be advisable to make a small change of course around mid-voyage. The alteration would only be of the order of a degree, but that would produce the required compensation over the distances concerned.

We have several times mentioned the use of radio and radar in space, and this is a subject of almost as much importance as the building of rocket motors themselves. It is true that without rockets space-travel would be impossible (at least in the present state of our knowledge)—but without radio it would also be severely handicapped. There would be no way of controlling our first unmanned missiles or of recovering information from them, and we would not be able to communicate with spaceships or send them navigational instructions.

Let us clear up at once two common misapprehensions about radio. First of all, like light and all the other radiations in the family of electromagnetic waves (infra-rad, ultra-violet, X-rays, gamma rays, and so on), radio waves will travel through the vacuum of space. This was proved experimentally on the astronomical scale long before the United States Signal Corps, in 1946, succeeded in getting radar echoes from the Moon. As early as 1932 Karl Jansky had detected radio "noise" from a source which he proved to be well outside the Solar System, since it showed no appreciable shift in the sky during the course of the Earth's annual movement. (It is believed that this radiation originates in the stars or nebulæ of the Milky Way, but its detailed investigation is now one of the biggest tasks of the new science of "radio astronomy".) Thus if we have sufficiently powerful transmitters, and sufficiently sensitive receivers, we could send radio signals of whatever kind we like (Morse code, sound, photographs, facsimile, vision, etc.) over any astronomical distance.

Many people have been somewhat confused by the existence of the Heaviside Layer or ionosphere, which as we mentioned in Chapter 2 reflects radio signals back to Earth and so makes long-distance services possible. This layer, however, reflects only waves which have a length greater than a certain critical value— a value which depends on a large number of factors but which is usually around 10 metres. *Radio waves shorter than this go straight through the ionosphere with little or no opposition and so can be used for interplanetary purposes.*

Our "window" into space will pass an enormous band of radio-frequencies, covering all the television and radar services, and in fact stretching, with only a few gaps, right down through the centimetre, millimetre and infra-red rays to visible light itself. Waves shorter than this (i.e. ultra-violet, X-rays and so on) cannot pass through the atmosphere as the air is opaque to them. (Only the longest rays of the Sun's ultra-violet spectrum get through the atmosphere and reach the Earth's surface. If they were not blocked far above our heads, life, as we know it, would be impossible on this planet, and evolution might have taken a different course.)

It seems safe to say, therefore, that astronautics will have all the wavelengths it will ever want to use, for this band of frequencies could carry something like a billion (1,000,000,000,000) speech circuits without interference! The question is, can we hope to send recognisable signals over such distances as those between the planets?

As far as our next-door neighbour, the Moon, is concerned, this can be answered at once. With quite a modest aerial system, even the feeblest of short-wave transmitters could be used to keep in touch with a lunar expedition. It would be a little more difficult to send television signals back to the Earth with the fairly low-powered transmitter a spaceship would be able to carry, but it could be done.

When true interplanetary ranges are considered, the power problem is naturally much more acute. Venus being, at her nearest, 100 times as far away as the Moon, we should need 10,000 times the power to provide the same kind of service with the same apparatus. However, sheer power is not the only answer. Most people are familiar with the large metal or wire-mesh structures, looking rather like glorified electric fires, which are used in some types of radar. These are simply "radio mirrors" for collecting incoming signals or focusing outgoing ones. By using such mirrors, and making them sufficiently large, we could increase range without putting up power.

It is reasonable to suppose that a spaceship could carry such a collector, folded up like an umbrella, with an effective area of a square yard or more. Very much larger mirrors could be employed at fixed stations on the Earth. (Indeed, some of up to 3,000 square yards in area have already been built for astronomical research!) By the use of such installations, there would be no great difficulty in sending speech as far as Mars and Venus, and code messages (which require much less power) could be sent at least as far as Jupiter.

The factor limiting radio range in space is probably the interference from the Milky Way, which has already been mentioned. This provides a background of radio noise which would swamp our own transmitters over great distances. Making the receivers more sensitive would merely bring in more of the "cosmic noise" and so would not improve matters. However, this does not seem to be a limitation which will worry us unduly if we are concerned only with sending messages over the relatively short distances found in the Solar System!

One limitation which *is* going to be a nuisance is that caused by the finite speed of radio waves. It takes $2\frac{1}{2}$ seconds for a signal to reach the Moon and back, travelling at the rate of 186,000 miles a second. If, therefore, one were talking to anyone on the Moon there would be this slight but annoying time-lag before there was any reply. (This point was brought out well in the film *Destination Moon*.) Over interplanetary distances this time-lag is so great that conversation would be quite impossible. It would take five and nine minutes respectively to get replies from Venus and Mars, even at their closest to the Earth!

This difficulty is a fundamental one, and is nothing to do with the inadequacies of our present-day equipment. Unless we discover something that can carry signals at a speed greater than light (which according to the Theory of Relativity is impossible) nobody on one planet will ever be able to converse, in the usual sense of the word, with anybody on another. One could listen to the voice of a friend on Mars, but the words that reached Earth would always have been spoken at least four minutes before.

It is worth mentioning that, for some purposes, the use of light waves might be more convenient than radio. A beam of light can easily be modulated by speech, and, in the heliograph, it was used for sending Morse long before radio was invented. By employing large searchlights as transmitters and using collecting mirrors passing the light to sensitive photoelectric cells, it should be possible to maintain communication over ranges of millions of miles. The limiting factor in this case would probably be the background light of the stars and nebulæ.

We have already mentioned the use of radar during the approach to a planet, and it will doubtless play an important rôle in probing the dense atmospheres of Venus and the giant planets when we send ships to survey them. The suggestion has often been made that spaceships should carry radar to give warning of approaching meteors, so that evasive action could be taken in time. A very brief investigation shows that this is about as practical as the White Knight's scheme for spiked anklets to protect his horse's feet from shark bites. Meteors are so tiny that no radar system could detect them more than a fraction of a second before impact—a warning period which would be quite useless. In any case, as we shall see on page 91, they are far too rare to be a serious danger to space-travel.

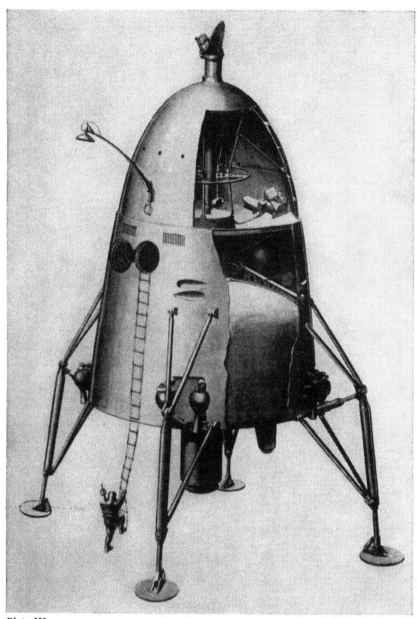

Drawing by R. A. Smith

LUNAR TYPE SPACESHIP: SECTIONAL VIEW

Plate IV

SPACESHIP ON MOON

Drawing by R. A. Smith

9. Life in the Spaceship

Down, down, down. Would the fall *never* come to an end?
LEWIS CARROLL—*Alice in Wonderland*

IN earlier chapters we have already mentioned a number of the technical problems which will have to be overcome before men can live and work without discomfort in spaceships. The two most important factors—air conditioning and temperature control—were discussed in Chapter 6. In addition we have to consider the provision of food and water, the natural hazards produced by dangerous radiations or meteors, and medical and psychological problems which might arise under these abnormal conditions.

Let us first examine the purely physical factors. A spaceship outside the atmosphere would be continually bombarded by the visible and invisible radiations from the Sun, and the mysterious and extremely penetrating "cosmic rays" which appear to reach us from all directions.

The only effect of the solar rays would be to warm the ship: they could not penetrate even the smallest thickness of hull. One precaution would be necessary to deal with them, for the dangerous ultra-violet rays, which could cause severe burns, would pass through the windows or portholes of the ship unless they were made of suitable material. Fortunately there are types of glass which are opaque to ultra-violet but will pass ordinary light.

Cosmic rays, on the other hand, cannot be stopped by anything short of a yard of lead, and the more powerful radiations will penetrate even that. Their origin and indeed their composition is still in doubt, but they appear to be charged particles (probably protons) travelling at speeds only a little less than that of light. When they enter the atmosphere, they produce a complex series

89

of secondary radiations as they smash their way through the air. The atmosphere thus acts partly as a shield, and partly as a source of additional rays, in much the same way as a brick wall, hit by a shell, would provide a large number of secondary missiles.

We pass all our lives under this bombardment, and there is no evidence that it does us any physical harm. At high altitudes the intensity of the radiation increases, reaching a maximum about twelve miles up, where it may be fifty times as intense as at sea level. With further increase of altitude the value drops again, reaching a uniform level about fifteen times that at the Earth's surface.

Although the effects of cosmic rays over prolonged periods of time, such as the weeks of flight on an interplanetary journey, are still unknown, it does not seem likely that they will present any serious danger. Pilots of high-altitude aircraft have spent many hours—in some cases, probably a total of many days—at levels in the atmosphere where cosmic radiation is at least as intense as in free space. Two men (Stevens and Anderson, in the stratosphere balloon *Explorer II,* 1935) went through the region of maximum intensity and spent several hours above it without ill effects. And it is worth recording that a number of *Drosophila* fruit-flies, the standard raw material of the geneticists, have been recovered safe and sound after a ride in a V.2—thus earning the distinction of being the first living creatures to travel outside the atmosphere! It seems, therefore, that the cosmic ray danger is just another of those mythical menaces which have threatened every new enterprise, and may be classed with the sea monsters against which Columbus was doubtless warned by his helpful friends.

Meteors, on the other hand, are certainly not mythical. On occasion they have punched holes a mile across in the Earth's crust, and during the course of every twenty-four hours the *total* number entering the atmosphere may reach, it has been calculated, something like 750,000,000,000,000,000. The vast majority of this stupendous number are far smaller than grains of sand:

only some 5,000,000 are as much as a tenth of an inch across, and *only five or ten are large enough to survive their passage through the Earth's atmosphere and reach the surface.*

When one compares the size of the Earth with that of any conceivable spaceship, the danger of meteors assumes its true proportions. Although at the speeds concerned (meteor velocities up to 160,000 m.p.h. may be encountered) particles only a few thousandths of an inch across could be dangerous, the chances of a penetration are negligibly small on journeys of moderate duration. Thus it has been calculated that, on the voyage to the Moon, there is less than one chance in 10,000 of encountering a meteor which could penetrate a steel sheet one-eighth of an inch thick.

It is interesting to imagine what would happen in a collision at such velocities—so much greater than any that are ever met on the Earth. One suggestion is that the meteor and the area of the hull at the point of impact would be instantly vaporised, with no penetration. In this case, a second shell an inch or so outside the spaceship's pressure chamber would act as a kind of bumper and provide a large measure of safety.

It should also be pointed out that, even if a meteor penetration did occur, in the majority of cases it would not be dangerous. If the pressure chamber itself were punctured by a meteor of moderate size, the resulting hole could be sealed by a suction pad before there was much loss of air. If it seemed worth while, self-sealing walls could be built on the principle employed in the construction of bullet-proof petrol tanks for aircraft, but it seems most unlikely that such drastic steps would be necessary.

It is certainly unnecessary to take any measures against the rare "giants", an inch or more in diameter, since an average-sized spaceship would have to wait several million years before a collision occurred with such a monster. The even bigger meteors are of course much rarer still.

At the other end of the scale, meteoric dust might be expected to produce, after a spaceship had been in operation for a considerable period of time, numerous tiny holes in the hull which would add to the natural air-loss from the structure. The magnitude of this effect can only be discovered by experiment, and its remedy may not be anything more difficult to apply than a coat of paint!

Meteors do not provide the only matter in space, though they are the smallest solid objects. In addition, spread between the planets and even the stars is an inconceivably tenuous gas, consisting almost entirely of hydrogen at a density some thousand million million millionth of that of air at sea level. For all practical purposes, this is a perfect vacuum: certainly the "interstellar gas" will be of no importance in travel between the planets, though as we shall see in Chapter 16 it might be significant when we begin to think of journeying to the stars. We mention it here because it raises again the question of "the temperature of space". These hydrogen atoms are moving at considerable speeds—speeds which in a gas under normal conditions would correspond to a temperature of 20,000 degrees F. or more. But because of the extremely low density of the "gas", the actual heat content is completely negligible and its presence would not affect a spaceship in the slightest. A few years ago a journalist, who had read that "space was at a temperature of several thousand degrees", wrote a sensational article on the theme that the Earth is surrounded by a belt of fire and interplanetary travel is therefore impossible. It would have been an interesting experiment to have immersed the gentleman concerned—with air supply, of course—in the 20,000 degrees F. interstellar gas without any other source of warmth. Before he became too numb with cold he might have learned to appreciate the difference between "temperature" and "heat"! For though the occasional hydrogen atoms might be at very high temperatures, one would have to gather them together over an enormous volume of space to collect enough heat to boil a thimbleful of water. A very similar state of affairs is encountered with the familiar indoor fireworks known as "sparklers". These eject brilliant stars which, though at temperatures high enough to be incandescent, are so very tiny that one can let them fall on the hand without any sensation of heat.

The same thing applies to the gases of the upper atmosphere, which are at temperatures of at least 3,000 degrees F. Although millions of times denser than the interstellar gas, even they are too rarefied to give any heat to a body immersed in them.

Before leaving this subject, it is interesting to calculate how much matter a volume of space equal to that of the Earth would contain. The answer is—about a quarter of an ounce of meteors, and two or three ounces of hydrogen! Perhaps these figures give, more clearly than any others, some impression of the emptiness of space.

We come now to the subject which has already been given the title of "space medicine", by analogy with the not-much-older science of aviation medicine. Although we may be able to solve the purely engineering problems of astronautics, unless we can overcome the physiological ones as well we will be doomed to remain for ever on our native planet, able to look at other worlds only through the eyes of our robots.

As far as temperature and pressure are concerned, we have seen that there is no fundamental difficulty in producing perfectly comfortable conditions. The provision of food also presents no problem apart from that of disposable payload. It is sometimes suggested that very concentrated foods may be developed in the future so that a whole meal could be taken, literally, in the form of a pill. Certainly this would be a boon to astronautics, though we have yet to see it recommended with enthusiasm in a French book on the subject. However, it is impracticable, as the body needs a certain amount of bulk or roughage, and though one can exist in a half-hearted sort of way on pills and tablets for many days, it cannot be regarded as a serious long-term proposition.

The daily food ration for polar explorers is two pounds, but it seems most unlikely that the occupants of a spaceship, living under conditions of zero gravity and thus doing practically no physical work, would have the same appetites as men pulling sledges over glaciers at sub-zero temperatures. The required food level in space might be well below one pound a day: nevertheless we will be pessimistic and fix the amount at the polar standard.

As we have seen on page 67, the oxygen consumption is also two pounds per day for man. The water requirements are not so easy to calculate, since the amount of liquid the body needs varies a great deal with circumstances. Moreover, unlike food and oxygen, water could be reclaimed almost completely (by distillation and purification) in the closed system represented by a spaceship, so apart from "topping up" to allow for the inevitable losses the actual consumption should be very small. If we assume it to be one pound a day, then the *total* amount of material consumed by each member of the crew is:

Five pounds a day.

This is certainly not an excessively large figure: for three men on the voyage to the Moon and back, with a week's stay there, the total consumption of food, oxygen and water would be about 250 pounds. On longer journeys the allowance could be reckoned at about three-quarters of a ton a year per man—a considerable but not impossible weight penalty.

It is probably fair to say that the only really abnormal conditions which the human body will have to cope with in astronautics are those arising from acceleration—or the lack of it. Most people who have read novels or seen films on the subject must now be firmly convinced that at take-off the crew of a spaceship would be subjected to an almost unendurable strain—their bodies possessing many times their normal weight owing to the ship's acceleration.

This is yet another of the difficulties of space-flight which in the past has been grossly exaggerated. At an acceleration of two gravities, it would take nine minutes to reach escape velocity:

at five gravities, only three and a half. Men can stand considerably higher accelerations for longer periods than this if they are suitably protected. But, in fact, the requirements of astronautics are nothing like as severe as these figures would suggest. There is no need to use very high accelerations during take-off, except for the first minute or so of powered flight.

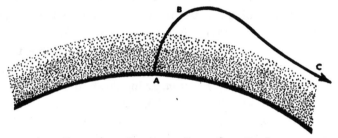

Figure 15. Departure Curve from Earth

The only point of employing high accelerations at all is to reduce the gravity loss caused by the Earth pulling the ship back. This would be a serious factor if the rocket climbed vertically at a low acceleration and so took a considerable time to reach escape velocity. However, if the ship could build up all its speed in the *horizontal* direction, gravity would not produce any loss of speed whatsoever and one could reach escape velocity in as leisurely a fashion as one pleased. This is clearly not practicable, but something not far from it would be achieved if the spaceship rose vertically for the first fifty miles and then veered eastwards, as shown in Figure 15. Gravity would then pull it downwards, but would not affect its horizontal speed, which would be steadily increasing under the thrust of the rockets. If the flight-programme had been correctly arranged, the ship would reach orbital velocity (say at the point C) before there was risk of it re-entering the atmosphere. It would then shut off its motors completely and wait to be refuelled, or it could continue to accelerate, as slowly as it wished, until it had reached escape velocity.

It will be seen, therefore, that only over the portion of the flight-path between A and B would a fairly high acceleration—perhaps as much as three gravities—be required. This would last just over a minute. For the rest of the powered trajectory, from B onwards, the acceleration need be little more than one gravity and the crew would feel no discomfort whatsoever.

Three gravities acceleration in the vertical direction means that the occupants of the spaceship would feel *four* times their normal weight—since gravity is adding its own force to that produced by the thrust of the rockets. Now it is possible for a man in normal health to stand such accelerations for two or three minutes *even in the sitting position*. Lying down, the tolerance to acceleration is very greatly increased and a man can survive even twenty gravities for periods of up to a minute. The human body is built, therefore, to far more stringent specifications than are ever likely to be needed for interplanetary flight.

Indeed, the design of the spaceship rather than that of its crew would determine the limiting acceleration that could be used, for large take-off accelerations would mean an unacceptable increase in structural weight to withstand them.

Four gravities in the "lying prone" position would not even be particularly uncomfortable if it lasted only for a minute. One could still move one's limbs and would not be entirely helpless. Thus raising an arm when under four gravities would demand about as much effort as is required when one lifts a heavy brief-case or a portable typewriter at arm's length—a feat frequently performed by office workers who can hardly be considered of exceptional physique!

In practice, the crew of a spaceship would have nothing to do during the ascent, for as we have already remarked it would be completely automatic.

It might be pointed out that, as the ship climbed and then veered eastwards, the apparent direction of gravity would remain unaltered. "Down" would still be from the nose of the ship to the tail, at least as long as the motors were operating. Thereafter, of course, there would be no "up" or "down" at all, for the reasons explained in Chapter 5.

No other operation would require such high accelerations as those met on the take-off from Earth. Landings and take-offs for bodies such as Mars and the Moon would be relatively mild affairs, and there appear to be no planets in the Solar System having gravities greater than that of our own world *and* possessing stable surfaces on which we should ever wish—or be able—to land. So much for the danger of excessive weight. But what of the other extreme—no weight at all?

Here we are confronted with a condition which we can never reproduce on the Earth, and so we are unable to make any tests before we encounter it. There is no difficulty in discovering how men will react to greater than normal weight: we simply have to put them in a centrifuge and whirl them round. But there is no way at all of *weakening* gravity. The only thing we can do is to look carefully at the functioning of the body, decide how gravity affects the various organs and physical processes, and then imagine what would happen if it were removed. The problem is essentially one of theoretical mechanics—the extremely complicated mechanics of the human body.

Some facts are obvious at once. In many respects our bodies show an almost complete indifference to the *direction* of gravity. We can eat, breathe, talk, think, use our hands just as well when standing as when sitting or lying. The only situations in which we feel definitely unhappy are those in which the feet are higher than the head—and even this condition of "negative gravity" can be tolerated for quite lengthy periods.

These facts show that gravity neither helps nor hinders most of the ordinary bodily processes. Invalids have lived for years in the horizontal position, where gravity, though of course it is present, can have very little effect on them—it cannot help them to swallow or digest their food, and the fact that the heart then has only a fraction of its normal work to do in lifting the weight of blood does not cause it to "race".

The consensus of medical opinion is that the only organ likely to be affected by the complete absence of gravity is the balancing mechanism of the inner ear. This contains a system of three semi-circular channels, partly filled with fluid and arranged at right angles. The movement of this fluid as the head turns is detected by suitable nerves, and as this is a purely inertia effect one would expect this particular function to be unaffected by the lack of gravity. In addition, however, there are two other organs in the inner ear which are believed to register gravity. These are the tiny cavities known as the "utricle" and "saccule" which contain microscopic solid particles called "otoliths" or, literally, "ear-stones". The otoliths may be regarded as carrying out the functions of plumb-bobs, detecting the direction of gravity and also measuring its magnitude.

It is an interesting and perhaps significant fact that people who for some reason are deprived of the use of these organs do not seem to suffer ill effects. They can also orientate themselves correctly even under water, when there is no sensation of weight to give any assistance. This is because vision alone is quite sufficient to give us a "reference system" without the aid of any other organs.

It therefore seems likely that the condition of weightlessness, though it may take some time to get used to, is not going to be harmful and need not even be unpleasant. (It may, in fact, be the reverse, for it will make possible at last the universal dream of levitation!)

The psychological rather than the physiological effects may be important, and it might be advisable to design the cabin of a spaceship to *look* as if up-and-down existed.

It seems quite likely that, after a prolonged period of weightlessness, the return of gravity might be a more uncomfortable experience than its cessation. Thus men who had been on a very long interplanetary journey would, when they returned to Earth, take some time to adjust themselves to the new conditions. They would not only feel very heavy and sluggish, but there would be an appreciable danger that they might treat vertical distances with insufficient respect. Much the same thing would apply to men returning to Earth after living on a planet of low gravity.

Finally, if it should prove essential or even desirable to avoid this condition of weightlessness, there is a fairly simple manner in which it could be done. There are indeed two, but only one is practicable. The inhabitants of a spaceship would have normal weight if the motors could maintain a constant acceleration of one gravity for the duration of the voyage—or, rather, an acceleration of this value until the half-way point and a *deceleration* thereafter, when the ship had been turned through 180 degrees. Since we have seen that it is barely on the verge of engineering possibility to carry enough fuel for accelerations lasting more than a few minutes, this solution can be dismissed at once. One day, when we are complete masters of atomic and perhaps even more powerful forms of energy, we may be able to do this sort of thing, but for a long time to come it will remain pure fantasy.

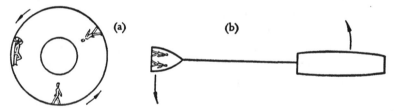

Figure 16. Producing "Artificial Gravity"

The other, and practical, solution involves the use of centrifugal force. Consider a spaceship of circular cross-section (Figure 16 (a)) spinning about its axis. Then, if there is no other force acting—that is, if the ship is in "free orbit"—it will seem to the crew that the circular wall is "down". They will be able to walk around on it and if the rate of rotation was correctly chosen they would have a sensation of normal weight. Indeed, there would be no way in which they could distinguish between this "artificial gravity" and the real thing.

There would be no impression of spinning, unless they looked out at the stars and saw them moving past the portholes. Even then the impression would be a purely psychological one—of the kind most people have experienced when sitting in a train and seeing another apparently moving on a parallel track. In those circumstances it is impossible, without some other evidence such as vibration or a check on an object which is definitely known to be stationary, to decide which vehicle is actually in motion. And, of course, there would be no vibration in a ship spinning on its axis in empty space, and no other reference system but the stars.

The rate of spin required to produce one gravity is quite small: a chamber ten feet in radius would have to make one revolution every three seconds. This spin could be imparted by small rocket jets of the type used for steering. They need be fired for only a few seconds to give the ship the necessary rotation, which would of course have to be removed again by another short burst of power before a landing was attempted.

Simple though this solution appears to be in principle, it does raise a number of problems. The psychological effects of seeing the people on the opposite wall standing "upside-down" might be unfortunate, though one could probably get used to it. In any case, it would be a simple matter to design the cabin so that one could not see the other side, though the crew would still have to get accustomed to the peculiar behaviour of the "floor" which was obviously curving upwards ahead of them and behind them, yet remained "flat" wherever one was standing. The convergence of verticals would also be noticeable: anyone standing a few feet away would be tilted towards one.

A man standing in a chamber of this size would have his head half-way to the centre, and so "gravity" would have only half the value here that it did at the wall. One would have to be lying against the wall to be in a substantially uniform field, but it seems unlikely that this effect would be objectionable or even very noticeable.

All the above difficulties would be overcome if the radius of the chamber were sufficiently large. A possible way of achieving this effect without making an impracticably large spaceship is shown in Figure 16 (b). If, when it had been launched on the correct orbit, the ship was separated into two sections which were connected by a cable and then set spinning, in the manner of a bolas, then the direction of "gravity" would be practically the same over the whole cabin, and it would not weaken towards the "roof". For a hundred-foot connecting cable, the period of revolution would have to be about eight seconds.

It remains to be seen whether it is worth while going to all these complications, and certainly it would make the spaceship designer's life much simpler if they prove to be unnecessary. For it must not be forgotten that the ship would have to be considerably stronger and hence heavier if it has to withstand these accelerational forces. In addition, special apparatus would be needed to take observations of the stars and to keep radio equipment aligned on Earth.

We have already touched several times on the psychological aspects of space-travel, and no doubt this question would be important on really long journeys. It is worth remembering, however, that properly selected men, united together in some common enterprise, can live and work together at close quarters for months or even years, as the records of Antarctic expeditions have shown.

We will return again to some of the medical problems of space-flight when we come to discuss the question of "space-suits" and planetary colonisation, but it seems appropriate to deal here with a point which, sooner or later, is bound to arise in practice despite all precautions. It is this—what would happen to the crew of a spaceship if a serious leak suddenly occurred and the air pressure dropped to zero?

This problem of "explosive decompression" is already of very great practical importance in high-altitude flying, since it may occur if a window or astrodome blows out (as indeed has happened more than once). The same thing could occur in a spaceship, with a complete loss of air pressure in a matter of seconds. This would be the case if a hole of area about a foot square was suddenly produced: an ordinary leak might take several minutes, or hours, to reduce the air pressure seriously.

It has often been suggested that the crew of a punctured spaceship would be killed instantly and their bodies perhaps ruptured by the expansion of internal gases when the pressure was released. This is not the case. Numerous tests have now been made of healthy human subjects and it has been found that a pressure drop of $7\frac{1}{2}$ pounds per square inch, occurring in less than half a second, can be tolerated. *This pressure change is twice that which would be produced if a spaceship, with a pure oxygen atmosphere, was suddenly opened to vacuum.* The experience is not even excessively uncomfortable and the subject remains fully conscious throughout it.

If the complete lack of oxygen continued, death would of course soon occur, but it seems probable that, in most cases, there would be time for emergency action before unconsciousness overtook the crew.

10. The Moon

Yet at first sight the crew were not pleased with the view
Which consisted of chasms and crags.

LEWIS CARROLL—*The Hunting of the Snark*

THE crossing of interplanetary space, though a technical problem which, as we have seen, will challenge all Man's ingenuity and resources, is not an end in itself but merely a beginning. There is no point in going to the planets unless we do something when we get there, and in the next three

Figure 17. Map of Moon

103

chapters we shall consider some of the practical difficulties which will have to be overcome before we can explore our neighbours in space and set up permanent colonies or bases upon them. We will begin with the Moon, logically enough, as it is closest to us not only in distance but also in time.

Since the invention of the telescope, more than ten generations of astronomers have spent appreciable portions of their lives (and astronomers are notoriously long-lived creatures) examining our satellite. Nowadays most of this work is done by amateur observers, for the professionals' interests lie a good deal further afield, among the stars and nebulæ. It is a somewhat surprising fact that the relatively small instruments that amateur astronomers can afford, or can make themselves, are not only adequate for this task but for many purposes can perform quite as well as the giant reflectors costing thousands of times as much. This is because one reaches the limit of practical magnifying power with a telescope of quite moderate size—that limit being usually set by the turbulence and irregularities of the atmosphere. Often, indeed, a large telescope may not perform as well as a smaller one, though on those rare occasions when visibility is virtually perfect, the larger the telescope the better. Anyone with sufficient energy and determination can make, for a few pounds, a telescope large enough to show more of the Moon's features than he could record in a lifetime.

Until this century, our knowledge of the Moon was largely descriptive, based chiefly on observations made by the eye—which far surpasses the photographic plate in its ability to record minute detail. Beautiful maps and drawings (those in Plate VII are fine examples) had given a good idea of the Moon's physical features, but the idea that we could ever discover anything, by actual measurement, about conditions on its surface would have seemed quite fantastic. Yet this is precisely what modern instruments can do. Temperature changes on the face of the Moon can be recorded by sensitive thermocouples used in conjunction with larger telescopes, and since the War a new field of research has been opened

Plate V

THE LUNAR BASE

Painting by Leslie Carr, based on a drawing by R. A. Smith

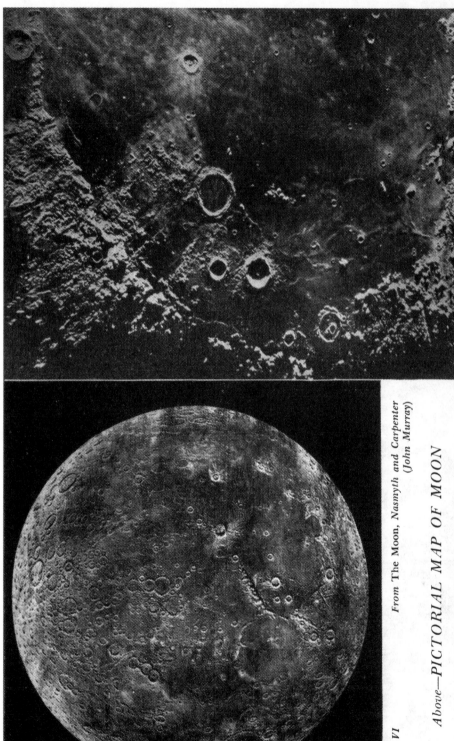

From The Moon, Nasmyth and Carpenter
(John Murray)

Above—PICTORIAL MAP OF MOON

Right—MARE IMBRIUM REGION
(Right) By courtesy of the Royal

up by the development of radio-astronomy. It is now possible
to measure the feeble radio waves *emitted* by the Moon—for all
heated bodies do emit radio waves, though at a power level
far too low to be detected in ordinary circumstances. From
such observations it has been possible to discover facts about the
nature of the Moon's surface, and the temperature below the
crust.

Perhaps the best impression of the Moon's geography (or, if
one wishes to be pedantic, selenography) is given by Plate VI
(a), which shows the general distribution of the craters and
mountain ranges. It should be explained that one can never see
all these formations at the same time, in the manner shown in
this drawing: indeed, when the Moon is full, its mountains cannot be seen at all, for no shadows are visible to produce an
impression of relief or solidity, and the disc appears flat and most
uninteresting. But everything shown in the plate can be observed,
with the very smallest of telescopes, if one watches the Moon wax
from new to full. The line of sunrise then marches steadily across
the disc, and one can watch the mountains climb out of the night
as their peaks catch the first rays of the Sun. Then the long black
shadows shorten until at noon they have vanished and even the
greatest mountains seem to have disappeared completely. In the
lunar afternoon they will be visible again as the shadows lengthen
towards night, and by watching these diurnal changes it is possible to learn a great deal about the shape and dimensions of the
Moon's surface features.

A key to the most important lunar formations is given in
Figure 17, which should be compared with Plate VI (a) and (b).
Over 670 features have been given names (mostly those of famous
philosophers, astronomers and scientists) and only a few of the
best known can be shown here. Many detailed maps and charts
of the Moon have been published, the most ambitious (that due
to the British observer H. P. Wilkins) being over eight feet in
diameter and showing thousands of formations.

Under perfect seeing conditions, an object only a few hundred

feet across could be observed on the Moon if it contrasted sufficiently well with its surroundings. Much smaller objects, such
as low ridges, can be detected by the huge shadows they throw
under low illuminations.

A good photograph, such as Plate VI (b), gives perhaps the
best impression of the Moon's surface, but it shows only a fraction of the detail which the trained eye can observe through a
moderate-sized telescope. Drawings such as those in Plate VII
are necessary to fill in the picture, and their production is a
challenge both to the observer's keen-sightedness and his artistic
ability.

Although the greater part of the Moon's visible surface is
extremely rugged, there are huge open areas (the so-called
"Maria"[1] or "Seas") which appear to be reasonably flat. We
cannot, of course, be sure of this until we have made surveys
from close at hand, but certainly there must be many places
where a landing could be made without difficulty. The radio
measurements mentioned above suggest that the lunar surface is
covered with a layer of dust (perhaps of meteoric origin) about
a millimetre thick. Some astronomers, before these measurements
were available, had advanced the theory that "dust oceans" hundreds of feet, or even miles, deep might be found on the Moon.
This would certainly have been a serious hazard to lunar explorers, but the evidence in favour of this depressing idea was
never very strong.

Although there are many mountain ranges on the Moon which
are quite similar to terrestrial formations—and of about the same
size—the characteristic lunar feature is the giant crater or ringed
plain. Some of these are up to 150 miles in diameter, and they
are quite unlike the normal volcanic craters found on the Earth.
This has resulted in a crop of scientific theories and a vigorous
controversy which has been going for at least a hundred years and
still shows no signs of abating. One school of astronomers insists

[1] Latin: plural of "Mare."

that the craters are caused by volcanic action, explaining their unusual features by the low gravity, the enormous tidal action of the Earth, and similar effects. Another powerful sect holds that they were caused by meteor bombardment in the remote past, while there are many dissenters who adhere to rival creeds. One sometimes wonders if the matter will be settled even when we *have* reached the Moon.

Another fairly common lunar formation is the "rill" or "cleft", of which several examples will be seen in Plate VII.[2] Some of these fissures are of great size and would be impressive, if not indeed terrifying, spectacles to an observer on the Moon. They appear to have been formed by "moonquakes" or similar rock movements, and are certainly not the result of water erosion as are some of the great canyons of Earth.

It is a rather common idea that the lunar mountains are much steeper than terrestrial ones, owing to the low gravity. This is quite incorrect. Although the gravity on the Moon is only a sixth of ours, this has no effect on the steepness of slopes. A pile of sand on the Moon would form a cone with the same angle as it would on Earth. However, the lunar formations may be more jagged and sharp, owing to the lack of normal erosion. Even this cannot be stated as a positive fact, for some "weathering" must have been caused by the great temperature changes between night and day.

Whether or not the Moon has an atmosphere is a question which is still under debate. No one doubts that any gaseous envelope it may possess must be far thinner than the Earth's, but *how* much thinner it is not easy to decide. Some tests indicate a value of about one-ten-thousandth of a terrestrial atmosphere, but other evidence points to a much lower limit—not more than a millionth. Since, in either case, this means that there is no question of life as we know it occurring on the Moon, it

---

[2] Both of these drawings show areas approximately 100 miles by 80 miles in extent. They were made by Mr. L. F. Ball, of the Lunar Section of the British Astronomical Association, using a 10-inch reflector.

might be thought that the discussion is of little practical importance.

This is not the case, for a rather curious reason. Owing to the weakness of the lunar gravity, if there were an atmosphere on the Moon its density would fall off with altitude much more slowly than does the Earth's. If Figure 2 were redrawn for the case of the Moon, the density lines would be about six times as far apart. Thus even if the density of the lunar atmosphere at ground level were only a ten-thousandth that of the Earth's, this would be amply sufficient to provide the Moon with an ionosphere which could reflect radio signals round the steeply curving horizon and, perhaps still more important, provide an even more effective barrier against meteors than our own atmosphere.

From time to time there have been reports of lunar mists which cause a temporary haze around certain craters and clefts. It is quite possible that very feeble volcanic action of some kind may occasionally take place and produce such clouds.

It is not safe, therefore, to say that the Moon is completely airless, nor can we be sure that it is completely waterless. Water cannot exist in the liquid state under such low pressure, it is true, but hoar-frost might reasonably be expected to form in the night as a temporary deposit. Some of the lunar peaks shine with an incredibly brilliant irridescence when they catch the first morning light, seeming to be mirrors rather than rock formations, and it is sometimes hard to dismiss the idea that they are not capped with ice. It is certainly quite likely that ice might occur in caves, where the temperature would be constant and far below freezing point.

Like Great Britain, the Moon has no climate—only weather. Because of the virtual lack of air, the temperature extremes between night and day are very great, and these are increased by the fact that the Moon turns so slowly on its axis that its day is 28 times as long as Earth's. Thermocouple measurements show that at midday exposed rocks directly facing the Sun reach temperatures just above that of boiling water (212 degrees F.). The temperature falls steadily towards sunset, dropping below zero while the Sun is still above the horizon. (Though any vertical rocks facing the Sun would still be extremely hot until the moment of sunset.) During the long night the temperature drops to about 250 degress below zero Fahrenheit, a total "daily" range of more than 400 degrees! For comparison, the highest and lowest temperatures ever recorded on Earth are $+136$ degrees and -94 degrees F.

It should be pointed out, moreover, that great variations of temperature could be encountered on the Moon simply by changing one's position slightly, say from a point in direct sunlight to one in shadow. So the values quoted above can be regarded only as indicating the extremes that might be met. Furthermore, they refer only to the surface layers and exposed rock faces of the Moon. Even a few inches underground, the variations of temperature would be greatly reduced owing to the insulating power of the lunar rocks. Deep inside the Moon, there would be no trace of them at all and the temperature would be constant at about -30 degrees F.

We have no information at all about the Moon's chemical composition: its density is considerably less than the Earth's but nothing can be deduced from this fact. It is reasonable to suppose that all the elements which occur on Earth will also exist on the Moon, though doubtless in different mineral forms.

To the question "Is there life on the Moon?" astronomers until quite recently would have returned an unequivocal "No", pointing out that the lack of air and the temperature extremes ruled out the possibility. Most astronomers would still maintain this position, but a considerable number of experienced observers claim to have detected changes in certain areas which suggest the existence of vegetation. It is by no means difficult to imagine that plant life might adapt itself to lunar conditions (some terrestrial cacti thrive in almost equally unpromising environments!), but for the present this must remain speculation.

One of the most curious—and tantalising—facts about the Moon is that, since it always keeps the same face turned towards us, we can never observe its other side. This absence of rotation relative to us means, of course, that the Earth must remain fixed in the lunar sky, never rising or setting. Owing to a slight "wobble" of the Moon as it travels in its orbit, the Earth would appear to move in a small circle around its average position, but this would hardly be noticeable, except at points near the visible edge of the Moon. Here there would be a narrow zone where the Earth would bob up and down on the horizon—and beyond this belt there would be the region where our planet is for ever invisible. For some reason this is often called the "dark side", but of course it would receive just as much sunlight as the rest of the Moon— nor is there the slightest reason to suppose that it will be very different in any other way.

When they learn what the Moon is like, most people find it very difficult to understand why anyone should want to go there—or what they could do when they arrived. The problem of lunar colonisation, and our satellite's extreme importance as an observatory, a scientific base, and a starting point for interplanetary expeditions, will be considered in our next chapter.

11. The Lunar Base

With the long shadows lying
Black, in a land alight
With a more luminous wonder
Than ever comes to our night.
LORD DUNSANY—*At the Time of the Full Moon*

TODAY we can no more predict what use mankind may make of the Moon than could Columbus have imagined the future of the continent he had discovered. Nevertheless, it is possible to foresee certain lines of development which appear likely as soon as we have reached the Moon, and we can also discuss, in general terms, the problem of making it habitable.

Some of the suggestions put forward in this chapter may well seem fantastic even to those whose imaginations have not been strained by what has gone before. Yet if we have learned one thing from the history of invention and discovery, it is that, in the long run—and often in the short one—the most daring prophecies seem laughably conservative. The things we are to discuss now will not happen quickly, and some will remain unaccomplished for centuries. But in the end they will be achieved—unless better and perhaps even more fantastic solutions are found.

The first lunar explorers will probably be mainly interested in the mineral resources of their new world, and upon these its future will very largely depend. If we are to set up permanent colonies, it is essential to discover oxygen, water, and materials from which food may be obtained. These are fairly long-term projects, but they will be very much in mind even from the beginning.

With the spaceships themselves acting as bases, the area around each landing-point will be carefully prospected by men wearing "space-suits". These have been described in fiction so often that there is some danger that we may take them for granted. Actually, they present technical challenges which are by no means fully solved. It is not simply a matter of building a slighty modified form of diving-suit and hoping that that will do—the problems involved are entirely different.

In the space-suit, because there is no pressure outside, there will be a very considerable force tending to make the suit rigid, like an inflated tyre, thus spread-eagling the unfortunate occupant. This difficulty does not arise in the case of the diving-suit, where the greater pressure is *outside*, but has already been encountered in the design of suits for high-altitude flight.

It is possible that, for this reason, space-suits may have to be fairly inflexible structures, with articulated joints at elbows, knees, etc. to allow for the necessary movements. They would thus resemble suits of armour, and might be just about as comfortable. Luckily the low gravity would simplify the problem of weight, though it must be remembered that the *inertia* would be unaltered. (On the Moon it would be six times easier to lift a sledge-hammer than on the Earth—but it would be just as hard to swing it.)

If even this sort of construction does not enable reasonably form-fitting space-suits to be practical, a more drastic solution may have to be found by making the suit completely inflexible—perhaps a simple cylinder with windows—and providing it with exterior metal limbs, the legs possibly motor-driven. Indeed, it must not be assumed that legs would be the best answer to the problem of locomotion. On a planet of low gravity a spring-operated pogo-stick arrangement might be much more useful!

The provision of air for such suits presents no great difficulty: oxygen for periods of up to twelve hours could be carried in cylinders of reasonable size. A more serious problem is temperature control, particularly on a body like the Moon, where if one moved a few yards from sunlight to shadow the temperature could fall four hundred degrees in a matter of seconds. It would be possible to insulate the suit from its surroundings—it could be a very efficient form of vacuum flask!—but this is highly undesirable, as the wearer would soon get heat-stroke owing to the thermal energy his body was developing. To counter this, it seems that space-suits would have to be provided with a small refrigerating unit for use during the day.

As there is no atmosphere—or certainly not enough to carry sounds—the explorers would have to keep in touch with each other, and with the ship, by radio. This of course presents no difficulty, but range would be rather short and the service somewhat erratic owing to the Moon's rugged nature and near horizon —only two miles away for a man six feet high. One of the first tasks of the expedition would probably be to erect as tall an aerial as possible, as is shown in Plate IV. A fifty-foot mast would give a radio range of about six miles, though this might be greatly extended if the Moon has an ionosphere.

Since a spaceship would hardly provide very commodious living quarters, an attempt would be made after the first few flights to build some kind of pressurised building or hut. This would probably be made of reinforced fabric or plastic material, blown up to form a hemispherical chamber and fitted with a small airlock. The whole thing would look very much like an Esquimo igloo, and would probably be silvered for heat insulation. Its inflation would mark the foundation of the Lunar Base.

For a considerable time all flights to the Moon would be directed to the same spot, so that material and stores could be accumulated where they would be most effective. There would be no scattering of resources over the Moon's twelve million square miles of surface—an area almost exactly the same as that of Africa. Where the first landing will be must of course be decided from the preliminary photographic surveys made by robot or piloted rockets: it will probably be in the Mare Imbrium or Sea of Rains (Plate VI (b), the largest of the great lunar plains. Certainly it will be on the visible side of the Moon, so that the expedition can keep in touch with Earth.

Within a few years of the first landing, it should be possible to establish a small camp on a permanent basis, keeping it supplied by a regular service of ships from Earth. A great effort would be made to set up an observatory with a telescope of moderate size: in fact it would be worth while building a spaceship for no other purpose than to carry a reflecting telescope of, say, twenty-inch aperture to the Moon. It could be here, in many fields of research, far surpass any terrestrial instrument in performance. The absence of an atmosphere would mean that it would always be possible to use it to the limit of magnification—and not on the few dozen occasions a year, at the very most, which is the best that can be expected on Earth. Since the lunar night lasts for fourteen of our terrestrial days, it would be possible to carry out many types of observation which cannot be performed on our rapidly spinning planet. Beneath the Moon's black and utterly cloudless skies, astronomers would at last be able to see the stars as they really are.

The Moon has so many advantages as an observatory that future generations may well wonder how we discovered anything about the heavens while we were still "earthbound". Under its perfect seeing conditions, such vexed problems as the existence of the Martian canals will be solved long before an expedition can reach that planet. There is, indeed, not a single branch of astronomy that would not benefit enormously.

This may be a good point at which to correct an almost universal fallacy—the idea that one would see the stars during the daytime on the Moon.[1] They would be there all right, because there is no atmosphere to swamp them with scattered sunlight. But the eye would not see them, because the intense glare from the brilliantly illuminated landscape would have made it too insensitive. To observe them, one would have to stand in shadow, shield the eyes completely from all sources of light, and wait a few minutes. Then they would become visible, first in tens and

[1] I am indebted to Dr. W. H. Steavenson for pointing this out.

THE LUNAR BASE 115

then in thousands—but they would vanish again as soon as one re-entered the sunlight.

When the time came to build really large telescopes, it would be a vastly simpler engineering proposition to do this on the Moon than on the Earth, since the low gravity would mean that a much lighter structure could be used. However, this is looking rather a long way ahead—to the time when there is a Lunar Colony and not merely a Lunar Base.

Astronomy is, of course, only one of the many sciences which will receive an immense impetus when the Moon is reached. When we come to the subject of "space-stations", or orbital bases, we will discuss in more detail some of the advances in the physical sciences, meteorology, radio and medicine which will be made possible by such conditions as low gravity and airlessness. The same arguments will apply, with minor modifications, to the case of the Moon.

We have already mentioned the quest for oxygen and water, and there is every reason to suppose that this would be successful. Although we cannot expect to discover appreciable oxygen in the free state, it should be remembered that this element comprises more than *half*, by weight, of the crust of our planet, and whatever materials the Moon may consist of will probably contain it in abundance. Given a sufficient source of power it can, therefore, in principle be extracted.

Water is also a common constituent of many minerals, and can be removed simply by heating. On the Moon it would be possible to obtain plenty of heat, during the daytime, by focusing sunlight with a simple arrangement of concave mirrors. However, the sheer physical problem of handling the quantities of rock involved would be considerable.

It would simplify matters greatly if water could be found in the free (frozen) state, and as has been pointed out it is quite possible that it may occur in caves, or on the surface in temporary deposits. Given electrical energy, it could then be electrolysed to provide oxygen, and the two major problems of life on the

Moon would be solved. We may safely assume that by the time lunar expeditions are feasible it will be possible to generate electrical power on a large scale from nuclear sources. The power unit of an atomic rocket would provide ample energy to run turbines and generators.

There is a good deal to be said for moving the lunar base underground at the earliest opportunity, if it proves possible to excavate the Moon's surface rocks without too much difficulty. An underground settlement would be easy to air- and temperature-condition, and its construction would not involve carrying materials from Earth. Possibly suitable caves or clefts might be found which could be adapted, but this is one of those matters about which speculation is fruitless at the present.

After air and water—perhaps even before—the search would be for material from which rocket fuel could be obtained. Water may again provide the answer. As was pointed out on page 78, it is a possible (though not the best) propellant for atomic rockets. Electrolysed to provide hydrogen and oxygen, it could drive conventional, chemical rockets. Once it became possible to refuel on the Moon, the whole picture would change and the economics of space-flight would improve by a factor of ten or more. It may well be necessary for this to happen before we can consider expeditions to the other planets (as opposed to mere orbital reconnaissances).

This point is so important that it is worth quoting a few actual figures. A spaceship leaving the Earth has to attain a speed of 26,000 m.p.h. in order for it to "coast" to Venus by the most economical route. If it left the Moon, the starting speed would have to be only 7,000 m.p.h. for exactly the same journey, because of our satellite's much lower gravity. If large supplies of fuel can ever be obtained on the Moon, it would become the key to the Solar System. Spaceships making any interplanetary journey would, on departure and arrival, refuel there. They would probably not land but would orbit the Moon while specially developed short-range ferry rockets brought fuel up to them.

It would even be good economics to refuel, *from the Moon,*
spaceships that had just reached orbital velocity *around the Earth*
and were circling it outside the atmosphere. Carrying rocket
fuel the quarter of a million miles from the Moon would be
cheaper than lifting it the few hundred miles up from Earth!

Another possibility is worth mentioning here. Because the
Moon has no atmosphere, the idea of the "space-gun", which
was ruled out for Earth, comes back into the picture. The
relatively low escape velocity (5,200 m.p.h. as against Earth's
25,000 m.p.h.) also makes this scheme much more attractive.
Such a device would not be a gun in the usual sense of the word,
but a horizontal or gently rising launching track, probably
operated electrically.

If one wished to project anything away from the Moon, it
would be much more economical to do this by means of a fixed,
ground installation (which could also have no restrictions on
weight) than by any kind of rocket device. It would be impracticable to use such a launcher for manned spaceships, because the
restricted accelerations needed would mean that it would have
to be about a hundred miles long. But fuel containers, if properly
designed, could withstand accelerations of perhaps a hundred
gravities and could be hurled into space by a launcher only two
miles long. They would then be intercepted when they were at
their furthest point from the Moon and travelling quite slowly
with respect to spaceships circling in an orbit at this distance.
(The interception presents some interesting problems which
might be solved in several ways: we will leave this as an "exercise" for the reader!) If, as would probably be the case, the fuel
containers were launched at less than escape velocity the empty
tanks would, some hours later, crash on the Moon again. It would
be easy to calculate the impact point, so this cannot be regarded
as a serious disadvantage. There is plenty of room on the Moon
for the returned empties to come down without adding noticeably
to the number of existing craters!

The provision of food for the colony is a complex problem

which can only be touched upon here. There are two obvious
lines of attack—pure chemical synthesis, or hydroponic farming
(see page 119). Both solutions are probably practicable, or will
be by the time the need for them arises.

It is not even totally impossible that, provided suitable soil
can be found or manufactured, plants could be developed—
perhaps from the native lunar vegetation, if this indeed exists—
which would flourish unprotected on the Moon's surface.

Until we know the answers to such questions as these, we
cannot say whether the future population of the Moon is likely
to be a few score scientists—or millions of men living comfortable
and, to them, quite normal lives in huge, totally enclosed cities.
The greatest technical achievements of the next few centuries
may well be in the field of what could be called "planetary
engineering"—the reshaping of other worlds to suit human
needs. Given power and knowledge (wisdom is rather useful, too)
nothing that does not infringe the laws of Nature need be
regarded as impossible. We will return to this theme when we
discuss the other planets, but it will already be apparent that the
conquest of the Moon will be the necessary and inevitable
prelude to remoter and still more ambitious projects. Upon our
own satellite, with Earth close at hand to help, we will learn the
skills and techniques which may one day bring life to worlds as
far apart as Mercury and Pluto.

An attempt to depict the Lunar Base as it might be some
decades after its foundation is shown in Plate V. In the centre
are the main buildings of the Base—its residential area, as it were—
surmounted by the observatory. Some miles away, on the left, is
the space-port with its associated services, fuel-storage tanks and
so on. In the left foreground the track of an electromagnetic
launcher is being laid.

The object in the middle distance looking like an oversized
electric fire is a solar power plant, collecting the sun's rays and
focusing them on to boiler pipes which carry their heated
contents to a turbine. The whole apparatus is pivoted so that it

can follow the Sun around the sky. On the Moon, where unbroken sunlight is received for 14 terrestrial days, such a power plant would be a very attractive proposition. (For some years the Russians have been developing similar engines for use in the Arctic, where there is no alternative source of energy, and, during the summer, sunlight is continuous for many weeks.) Even when nuclear power is cheaper and more convenient than solar energy, such plants might still be retained as "stand-bys".

The green tubes surrounding the Base are the pressurised glasshouses of the hydroponic farms. This method of cultivation (sometimes known as "soil-less farming") would seem well suited to lunar conditions, where enclosed space would be at a premium. The plants are simply supported on netting over tanks through which nutrient liquids flow, and if plenty of sunlight is available large crops can be grown very rapidly.

Since conventional flying machines obviously cannot operate on the airless Moon, long-range lunar transport presents something of a problem. Rockets would almost certainly be uneconomic for short distances, and it appears quite likely that railways would be extensively used. Pressurised vehicles with large balloon tyres would also be employed for much the same duties that they fulfil on Earth. Their motors would be electric, operated by storage batteries, or else turbines, driven by reacting rocket fuels, either directly as in a gas turbine, or indirectly through the use of some intermediate fluid.

The Lunar Base depicted here may seem somewhat impressive, not to say fantastic, but it must be remembered that a full century may lie between it and the scene in Plate IV. That length of time, with the aid of atomic power, would be amply long enough to change the face of a world.

12. The Inner Planets

Hesper—Venus—were we native to that splendour
or in Mars,
We should see the Globe we groan in, fairest of
their evening stars.

Could we dream of wars and carnage, craft and
madness, lust and spite,
Roaring London, raving Paris, in that point of
peaceful light?

TENNYSON.

WE will now complete the survey of the Solar System which we began in Chapter 2, concentrating this time on the physical conditions of the planets rather than their dimensions and distances apart—though it is worth looking at Figure 1 again to remind ourselves of these.

For reasons which are still unknown, the planets appear to form two quite distinct classes. On the one hand are the relatively small solid bodies like the Earth, Mercury, Mars, Venus and Pluto. These range in size from eight thousand miles in diameter downwards, and all have a density several times that of water. They can properly be called "terrestrial type" planets, and they probably consist of much the same materials as does the Earth (though it may be unsafe to generalise about Pluto, concerning which practically nothing is known except its diameter).

On the other hand are the giant planets—Jupiter, Saturn, Uranus and Neptune. The smallest of these has four times the diameter of Earth, but their densities are extremely low (in the case of Saturn, being actually *less* than that of water). We are forced to conclude from this that the four giant planets are partly gaseous or liquid, perhaps possessing solid cores only at great

Plate VII

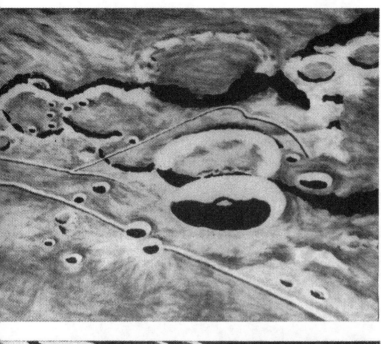

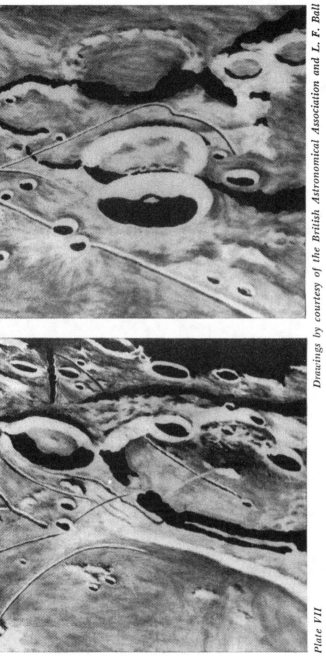

Drawings by courtesy of the British Astronomical Association and L. F. Ball

LUNAR FORMATIONS

Hevel and Lohrmann

Sirsalis

Plate VIII

From the film Destination Moon, by courtesy of Eagle-Lion

THE SPACE-SUIT

depths below an immensely thick atmosphere. Telescopic observations support this theory, for Jupiter and Saturn show changes—such as occasional vast disturbances—which could hardly happen if they were solid bodies.

In addition to the planets, the Solar System contains some thousands of asteroids or minor planets, whose diameters range from 480 miles down to, in all probability, a few yards. They occupy a broad, diffuse belt between Mars and Jupiter, but some wander as far afield as Saturn and others go sunwards to Mercury. Their orbits are frequently very eccentric and highly inclined to that of the Earth. This characteristic is also shared by the last category of wanderers round the Sun—the comets, those huge, tenuous masses of gas which for thousands of years have terrified mankind by their portents in the sky. Many are associated in some way with the meteor swarms which sweep through space, invisible until they enter the Earth's atmosphere.

But let us look first at the planets, considering them in the order, as far as can be foreseen today, in which they will be visited. As we have already seen, there is little to choose between Mars and Venus in this respect. It is quicker to reach Venus, but takes rather more power—at least if a landing is to be made.

The case of Venus, sometimes called Earth's sister planet, is a tantalising one. As at her closest she is only 25,000,000 miles away, it might be expected that we would know a great deal about her. Actually, we know practically nothing—not even the fundamental fact of her rotation period, which we know for all the other planets except remote and tiny Pluto.

The resemblance between our world and Venus virtually begins and ends with size. Venus is only about 4 per cent. smaller than Earth, and has a gravity about 10 per cent. weaker—a reduction we should probably not notice. The fact that she is considerably nearer the Sun means that her average temperature is a good deal higher than Earth's, but the polar regions and a broad belt in high latitudes would be perfectly tolerable to us—other things being equal.

However, they are not. In the telescope Venus appears as a brilliant, silvery crescent, revealing at the most only faint and fugitive markings, but more usually showing none at all. There are none of the mountains and valleys which make the Moon of such absorbing interest, nor any of the polar caps and deserts which we can see on Mars. The face of Venus is utterly featureless —or, to be more accurate, it is for ever hidden from us by clouds. These clouds make her the most brilliant object in the sky— so brilliant that she can be seen with ease in daytime if one knows exactly where to look—but they must cut Venus off from the glory of the stars and even the full light of the Sun. It is possible that direct sunlight may never reach the surface of the planet, so that even at noon the illumination is no more than a kind of submarine twilight.

One of the most mysterious things about an already sufficiently enigmatic planet is that the Venusian clouds are apparently not composed of water. No trace of water or oxygen has been detected by the spectroscope, the only gas which has been discovered in the atmosphere of Venus being carbon dioxide. This is present in enormous quantities: there is probably more than a thousand times as much of it above the clouds of Venus as in the whole of our atmosphere.

What lies below the clouds we can only guess. Since carbon dioxide is one of the heaviest of the common gases, there is little hope that oxygen might be found at lower levels. Being a lighter gas, it would all have floated to the top.

It has been suggested that the clouds really consist of dust, stirred up by continual storms raging between the hot and cold sides of the planet. One objection to this plausible but not very heerful idea is that, if this were true, one would hardly expect louds of such dazzling whiteness—they would surely be brown or dirty grey.

The fact that Venus turns slowly on her axis is proved by measurements of the heat emitted by the planet's night side. This is at a much higher temperature than it would be if she always kept one face towards the Sun. On the other hand, her rotation period is a good deal longer than Earth's, because if the planet were spinning at all rapidly the spectroscope would have revealed it. The Venusian day is therefore at least ten and probably more like twenty times as long as ours.

It will be seen, from the above facts, that there is no foundation at all for the rather common idea that Venus is a world of oceans and steaming swamps, like our own planet some thousand million years ago, and perhaps inhabited by primitive forms of life. Oceans seem very unlikely, as we would have detected the resulting water vapour in the atmosphere. There is probably no vegetation, at least of the terrestrial type, as growing plants release oxygen and no trace of this has been detected. Venus may be a planet in such an early state of evolution that life has not had a chance to develop—although this seems contrary to what one would expect from theory. All the planets appear to have been formed at about the same time, and as Venus is slightly smaller than the Earth one would imagine her to be ahead of us, not behind us, on the evolutionary path. In that case it would be more reasonable to anticipate much more advanced life-forms than on Earth, though possibly of a completely alien type, rather than creatures that might come from our own primeval past.

If intelligent beings exist on Venus (and our lack of knowledge makes detailed speculation on the subject completely worthless) they have probably had quite a different history of scientific development. On Earth, astronomy was the first of the sciences: observation of the heavens taught Man that law and order existed in at least part of Nature, and from this beginning he went on to find them everywhere. If the Venusian clouds are so thick and so permanent that the inhabitants can never look out into space, they must have developed the other sciences before they came to astronomy. The study of the stars would not begin until the Venusians had developed flying machines which could take them above the clouds, or radio equipment which could pick up the waves which, as we ourselves have only just discovered, come from the Milky Way. If our hypothetical Venusians had to rely on radio as their only astronomical aid, their picture of the Universe might well be a curious one. They would know a good deal about the structure of the Milky Way—though they might not be able to guess what its nature actually was—and they would know the shape, size and possibly even the distance of the hidden Sun. But they might never have guessed at the existence of the other planets, which do not emit radio waves.

Or which, to be more accurate, did not emit radio waves until recently. . . . Some time in the early 1940's, if the Venusians possessed radio receivers working at wavelengths of a few centimetres, they would have picked up the transmissions of the first high-powered microwave radar sets when their beams happened to sweep in the right direction. One can imagine the excitement of the Venusian scientists at the discovery of this strange source of radiation so close at hand. That it was relatively close they would have realised after a few days' observations, for it would be shifting rapidly against the fixed background of the Milky Way. After several months they would know that the source was travelling round the Sun in an orbit similar to that of their own planet, but of greater radius, and this might be their first glimpse of the fact that theirs was not the only world in space.

Venus may present a very difficult problem for the first astronauts to reach her, and it may well be many years before any attempt is made to descend through her clouds. Before this happens there will certainly be detailed radar surveys of the hidden surface. Such surveys, carried out by orbiting spaceships, would show the fine details of coastlines, if there were any, and such features as lakes and rivers. Even if Venus is a wind-swept desert with no outstanding geographical markings, it will be possible to obtain an accurate contour map by radar.

Venus has no satellite, a deprivation shared only by Mercury and possibly Pluto. If there are seas on the planet they would not, however, be tideless, for the Sun would be a more potent raiser

of tides in the Venusian oceans than is the Moon in the seas of
Earth. Since the location of Venus' polar axis is unknown, we
cannot say if the planet has seasons as we do on Earth. If they
occur, they would be considerably shorter than ours, since the
length of the Venusian year is only 225 days.

It is, perhaps, a relief to abandon this exasperating planet and
try our luck elsewhere. In the case of Mars, we are not confronted
with impenetrable clouds: we can see the actual surface of the
planet and can make maps of its main features. Moreover, when
it is nearest to us Mars turns its illuminated face full towards the
Earth—unlike Venus, who passes between us and the Sun on such
occasions and is thus completely invisible.

Despite these advantages, our knowledge of the planet is full
of gaps and there are rival interpretations even of the admitted
facts. Because of its distance, an observer of Mars, using a large
telescope under conditions of good seeing, is in much the position
as someone looking at the Moon with the naked eye—or, at the
best, with a pair of weak opera-glasses. Though we have telescopes
that could bring Mars to within a tenth of the Moon's distance,
it is impracticable to use such magnifications, because our atmosphere is not steady enough. Simply increasing the power
of a telescope very soon ceases to show any finer detail and
in fact soon shows less: it is like looking at the reproduction
of a photograph in a newspaper through a magnifying glass—
the greater power only reveals the "graininess" of the image. To
aggravate matters, as the orbit of Mars is notably eccentric,
really close approaches of the planet occur at rather rare intervals
—the best approaches of all being 15 years apart. (The next will
be in 1956.)

Let us first consider the undisputed facts about our little
neighbour. It is just over half the size of the Earth (4,200 miles
in diameter) and thus its surface area is 25 per cent. of Earth's.
But three-quarters of our world is covered with water, and since
there are no oceans on Mars it follows that its land area is just
about equal to Earth's.

The Martian day is very nearly the same length as the terrestrial one, being only half an hour longer, and the axial tilt
of the planet is also almost the same as Earth's. Mars therefore
has seasons just as our planet has, but since the year lasts 687
days they are nearly twice as long. The changing seasons, as we
shall see later, produce important effects which can be observed
even across the millions of miles of space which separate us from
the planet.

In the telescope Mars shows three main types of surface mark-
ing. Most prominent are the brilliant polar caps, which wax and
wane alternately in the two hemispheres, almost disappearing
in summer and coming half-way down to the equator in winter.
Not so bright, but still very prominent, are the red or orange
areas which cover most of the planet. Finally there are the
irregular, dark regions which form a belt around Mars roughly
parallel to the equator. (See Plate X.)

These are the permanent markings. In addition, temporary
clouds and haziness can sometimes be observed, proving that the
planet has an extensive atmosphere.

The behaviour of the polar caps immediately suggests that
they are composed of ice, and this explanation is now universally
accepted. The Martian ice caps, however, must be far thinner
than the enormously thick and permanent crusts which lie at our
poles. This is obvious from the fact that even the mild summers
of Mars are warm enough to make them shrink so much that on
occasions the southern cap vanishes completely. They may therefore be only a few inches thick—the equivalent of a light fall of
snow, in fact.

The orange regions which give the planet its characteristic
colour show no seasonal changes, and are generally considered
to be deserts. This word must not, however, conjure up a picture
of a drab, sand-covered waste. The Martian deserts show some
extremely brilliant colouring—brick-red and ochre being the terms
frequently used to describe them. They may resemble some of
the incredibly spectacular and garish deserts of Arizona. It has

been suggested that the planet's characteristic redness is due to the presence of metallic oxides, particularly iron oxide. If this is the case, Mars is a world which has, literally, rusted away.

The Martian deserts are probably fairly flat, for we should be able to detect any high mountains by the irregularities they would cause on the line between night and day. There is, however, no reason why hills or plateaux a mile or two in height should not exist, and indeed there is some evidence for mountains near the South Pole. The ice-cap occasionally splits into two sections as it shrinks, leaving an isolated white patch which is always at the same location, as might be expected to happen if there were high ground there.

Undoubtedly the most interesting areas of Mars are the dark regions, which show seasonal changes linked with the melting of the polar caps. The early observers made the fairly natural assumption that these regions were seas, and christened them "Maria". The history of lunar nomenclature was thus repeated, and though we now know that Mars is as bereft of seas as is the Moon, the names are still used. (Mare Cimmerium, Mare Serpentis, Mare Sirenium being among the more fanciful inventions.)

With the melting of the polar caps in the spring and early summer, a belt of darkness spreads slowly down towards the equator across the "seas". This change is so obviously produced by the release of water from the caps that the evidence for the growth of vegetation is overwhelming. (It is, of course, conceivable that the change might be due to chemical reactions among mineral deposits of some kind, but there seems little point in advancing this complicated explanation in place of the obvious and simpler one.) The colour changes which occur are strikingly similar to those which we should witness if we observed our own Earth from space. During most of the Martian year, the "Maria" are blue-green or blue, but in the late winter and early spring they become chocolate brown.

Before we jump to any conclusions regarding life on Mars, we must consider what is known about its atmosphere. The facts revealed by the spectroscope are rather disconcerting—there is no sign of oxygen, and we have tests which could detect the presence of this gas even if it were only a thousandth as common as in our atmosphere. Carbon dioxide has been observed: it is about twice as abundant on Mars as on Earth. Water vapour has not been detected in the atmosphere, but infra-red bands due to ice have been observed in the polar caps.

The air pressure at the surface of Mars is certainly very low—perhaps a fifteenth of its sea-level value here. We would have to ascend higher than the summit of Mount Everest to encounter so low a pressure on Earth, and even if the Martian atmosphere consisted entirely of pure oxygen we could not survive in it. The fact that it contains virtually no oxygen at all rules out any forms of animal life resembling those on Earth. It is probable that the bulk of the atmosphere consists of inert gases such as nitrogen or argon.

Although the pressure is so low the Martian atmosphere is very deep: the weak gravity (one-third of Earth's) means a much slower falling-off of density with height than on Earth. This is supported by the fact that clouds have been observed as much as 20 miles above the surface of Mars.

Despite the tenuous nature of the Martian atmosphere, it is surprisingly hazy, and normally blocks out the light towards the blue end of the spectrum. The reason for this is unknown: although one is tempted to explain it by the presence of fine dust in the atmosphere, it is difficult to see how so thin a gas could support much solid material.

This particular mystery is less important than the undoubted fact that the Martian atmosphere is very tenuous and contains no oxygen. The oxygen has probably not been lost: it is still there, locked up in the deserts which cover so much of the planet.

Perhaps the Martian plants, if they really exist, can obtain the oxygen they need from the soil rather than from the atmosphere. It should be remembered that the other basic raw materials for plant life are carbon dioxide, water and sunlight, all of which are certainly present on Mars.

The prospects for animal life appear somewhat gloomy, unless evolution has produced creatures which do not require oxygen at all. Otherwise, conditions on the planet are not too unfavourable and even life-forms similar to some found on Earth might exist. The low pressure and the scarcity of water are obstacles which could be overcome by the techniques of biological engineering which Nature has already developed here.

Nor is the temperature of Mars so low, despite its greater distance from the Sun, that life would be severely handicapped. At noon during the summer, temperatures of 80 degrees F. have been recorded by the thermocouple, and the equatorial regions of the planet must be not much colder, on the average, than the temperate zones of Earth. However, the range of variation is much greater, the Martian nights and winters being extremely cold.

It is worth remarking that the seasonal variations would not be a great hardship to animal or mobile forms of life. Owing to the smallness of the planet, the length of the year and the absence of geographical barriers, it would be quite easy to migrate from one hemisphere to another with the changing seasons. The average speed required would be only five or ten miles a day. Presumably non-mobile forms of life would go into hibernation during the winter, as do the plants of our Antarctic.

The shortage of water is probably one of the greatest handicaps to Martian life, and the yearly melting of the polar ice is clearly of extreme importance. It is possible that the water is carried from the poles southwards in the form of vapour, not in the liquid state. Although the air pressure is sufficiently high for liquid water to exist on Mars, none has ever been observed. Bodies of water only a few hundred yards across could be detected by the reflection of sunlight from them—a feature which would be very obvious to any observer watching the Earth from space.

It is possible that temporary lakes might form around the polar caps during the spring, and indeed a dark fringe is then observed around the melting ice. We would not be able to see, from Earth, the reflected sunlight in lakes at such high latitudes, so the test mentioned above cannot be applied to these regions.

One is tempted to assume that the dark areas of the planet may be the beds of long-departed seas, at a lower level than the surrounding desert, but there is no direct evidence for this.

The planet's main geographical (or aerographical) markings are shown in Plate X, which is drawn on a Mercator projection. It should be explained that the details shown here have been collected over many years of patient observation, by many observers. The disc of the planet also shows a vast number of minute markings which can only be glimpsed at moments of perfect seeing and cannot be represented on any drawing.

This is, perhaps, the moment to say something about the much-discussed "canals"—the network of fine, narrow lines reported by Schiaparelli and Lowell towards the end of the last century. Lowell was convinced that the "canals" formed a vast irrigation system, built by an intelligent race to conserve its dwindling water supplies. Few astronomers today accept this interpretation, and most do not believe in the existence of the canals at all. Yet there can be little doubt that large numbers of curious linear markings do exist on the planet. Even if they do not actually form unbroken lines, many of them seem to be arranged in a rectilinear fashion—but this does not mean that they must be artificial. They could quite possibly be old river beds, canyons, or similar formations, and it is probably safe to say that nowadays few, if any, astronomers could be found who believe that there is the slightest evidence for intelligent life on Mars. If anyone finds this discouraging, we might point out that the Martians could hardly have detected intelligent life on Earth if their telescopes were no better than ours![1]

[1] This is perhaps debatable. They might be able to see the lights of our great cities shining on the dark side of the planet.   •

Mars has two tiny satellites, only ten or twenty miles in diameter. Phobos, the closer of the two, is so near the planet that it is invisible from the polar regions, being hidden by the curvature of the globe. As it moves round Mars more quickly than Mars revolves on its axis, it rises in the west and sets in the east. Much of the time it must be eclipsed by the shadow of the planet, and as it would be about a quarter the apparent size of our Moon it would provide only a few per cent. of its light. Deimos, the outer satellite, is very much less conspicuous still and may not even show a visible disc to an observer on the planet —being perhaps merely a bright star.

These tiny Moons may well be the first extra-terrestrial bodies, next to our own satellite, on which human beings will ever land. Since their gravitational fields are negligible it would take very little power for a spaceship to make contact with them once it had entered an orbit round Mars. The gravity of Deimos must be so low that a man could jump clear away from it—reaching escape velocity with his unaided muscles!

This summarises what is known and what may be reasonably conjectured about Mars. The probability that it has plant life of some kind is very high, though there is no evidence at all for animal life or intelligence. One can imagine these if one postulates the existence of beings who do not require gaseous oxygen, or who have reached the fairly modest level of scientific achievement necessary to build pressurised, oxygenated cities. Here, of course, we enter the realm of pure speculation and one person's guess is as good as another's. It is advisable not to be too dogmatic about Mars at the present time, for when the next close approach of the planet occurs in 1956, the remarkable new telescopes and techniques developed since the last favourable opposition may produce some real advances in our knowledge. In particular, the 200-inch telescope should be able to provide unambiguous photographic evidence for the existence—or otherwise—of the canals.

Mars and Venus are the only two planets on which there was the slightest hope of finding terrestrial-type forms of life, and are also the only planets on which we could survive with relatively simple mechanical aids. (We will discuss the problem of exploring them, and the other planets, in Chapter 14.) As we go further afield, we come across worlds so much more alien that even our two not-very-promising neighbours seem paradises. It is possible that after Mars and Venus have been reached, the next planet to be attained will be Mercury, the world closest to the Sun. It is somewhat smaller than Mars, and appears to have a very tenuous atmosphere. In many respects the system Sun-Mercury resembles the system Earth-Moon, for Mercury always keeps the same face to the Sun although, just as the Moon does, it rocks to and fro slightly during each circuit. One side of the planet is thus bathed in perpetual and torrid sunlight, while the other is in eternal night. Between these regions is a sort of twilight zone in which the Sun bobs up and down on the horizon and something analogous to day and night exists.

The centre of the disc which faces the Sun is of course intensely hot—probably at over 700 degrees F., so that lead and tin would melt here. Towards the edge of the sunlit hemisphere the temperature falls steadily and there is a fairly wide belt whose exploration would present no great difficulty, on this score at least, as it would be little hotter than Earth. Over the edge of the planet, on the night side, it would be incredibly cold, for no heat could reach here except by conduction through the solid rock. It has been suggested that on Mercury will be both the coldest and the hottest points in the Solar System: the night temperature may be as low as —450 degrees F.—not far short of the absolute zero.

The surface of Mercury is probably very similar to that of the Moon, but its distance and proximity to the Sun make it impossible to observe more than a few indistinct markings which show a curious resemblance to the "canals" of Mars. However, as far as we know it has never been suggested that there is a race on Mercury irrigating its crops with molten lead!

We must now leave the neighbourhood of the Sun and travel out beyond Mars, to the lonely giant planets and their great families of satellites—each a miniature Solar System in itself. Although Mercury, Mars and Venus were each in their way very different from Earth, they had certain points of similarity: they were cousins, even if the relationship was no closer. But the planets we are to visit now belong to a different family—almost, indeed, to a different species.

13. The Outer Planets

Up from Earth's Centre, through the Seventh Gate
I rose, and on the Throne of Saturn sate.
EDWARD FITZGERALD—*The Rubáiyát of Omar Khayyám*

AS Figure 1 shows, beyond Mars the scale of the Solar System widens rapidly. Between Mars and Jupiter there is what seemed, for a long time, to be a disproportionately great gulf, as if a planet had been overlooked.

At the end of the 18th century, an attempt was made to locate this missing world. The result of the search was unexpected: not *one* planet was found, but hundreds—and we are still nowhere near the end of them. Known as the asteroids (an unfortunate word, as it means literally "little stars") the largest of these worlds is only 480 miles in diameter, and the smallest we can detect cannot be more than a mile or two across. They have orbits of every possible size, eccentricity and inclination: some range out as far as Saturn, and at least one goes nearer the Sun than Mercury. But the majority stay in the zone between Mars and Jupiter.

The total number of asteroids, of all sizes, must run into at least five figures. Until recently, astronomers were unable to keep track of even the 1,500 or so already detected—the work involved in calculating their orbits was too great. With the modern development of electronic computers, this difficulty has been overcome and almanacs for minor planets can be calculated and printed quite automatically. With this tedious work taken off their hands, astronomers no longer regard the asteroids with quite such a jaundiced eye.

Even the largest of these little worlds, Ceres, is far too small to possess an atmosphere—its gravitational field is so weak that any gas would escape into space immediately. Nothing whatsoever is known about their physical composition or surface features, since the vast majority appear simply as dimensionless points of light in the telescope. The smaller asteroids are probably not even spherical, but are simply jagged lumps of rock—mountains wandering through space.

Whether they will be of any interest to astronautics, only the future can tell. Although so many thousands of them exist, they cannot constitute a "menace to navigation", as has sometimes been suggested. The gulf between Mars and Jupiter is too enormous for a few thousand, or even a few million, asteroids to go very far towards filling it.

It is convenient to treat Jupiter, Saturn, Uranus and Neptune together, for they differ in degree rather than in kind. All have these points in common: they have a very low density, have atmospheres composed of the light gases hydrogen, methane and ammonia, turn very rapidly on their axes, and are extremely cold. Jupiter, being the nearest and also the largest, is the most easily observed of the four: much of the information gained about him probably applies to Saturn, Uranus and Neptune.

We can see no permanent surface markings on these planets: what we observe is the top of an immensely deep and turbulent atmosphere, perhaps thousands of miles thick. They may indeed possess no solid cores: the compressed gases may go on getting denser and denser until the centre of the planet is reached, with no definite transition from gas to liquid.

Even Jupiter, the warmest of the giants, is at a temperature of below —200 degrees F. The physical conditions of these weird planets would not only seem to preclude all possibility of life, but would also appear to rule out any direct exploration. No doubt spaceships orbiting a few thousand miles out may one day probe those immense atmospheres with radar beams and other tools of future science, but any closer contact than this is difficult to imagine. Yet it would be very unwise to rule it out as a long-term possibility.

It is on the twenty-seven satellites of the giant planets that

the explorers of the next century may find landing-places. Five
of these worlds are larger than our own Moon, and Titan, the
sixth satellite of Saturn, is almost as big as Mars. It has the distinction of being the only moon with an atmosphere of its own,
the atmosphere being one of methane (better known as "marsh-
gas" or "fire-damp").

Four of Jupiter's moons have been named—Io, Europa, Ganymede, Callisto—but thereafter the astronomers gave up and
simply numbered the rest V to XI. The smallest of all (X and XI)
are only about 20 miles in diameter, and are probably the faintest
planetary objects ever detected.

It is an interesting fact that although the four large satellites all
lie within little more than a million miles of Jupiter, travelling
between them would involve almost as much power as does a
journey from Earth to Mars or Venus. This, of course, is a
consequence of Jupiter's extremely powerful gravitational field.
Some idea of its magnitude is given by the fact that escape velocity
at the planet's surface is 130,000 m.p.h., compared with the
Earth's mere 25,000!

Although nine moons of Saturn have been discovered, there
can be little doubt that others still remain undetected. The fact
that Jupiter's smallest satellites are very much fainter and
smaller than Saturn's rather suggests that, for the outer planet,
the search has not been as thorough as it might be! It is certainly
a little odd that Jupiter has five moons less than 40 miles across,
yet Saturn's smallest is 200 miles in diameter. These have all been
dignified with names, not merely with numbers, and the roll-call
is so poetic that we cannot resist giving it in full (working
outwards from the planet): Mimas, Enceladus, Tethys, Dione,
Rhea, Titan, Hyperion, Iapetus and Phoebe.

All the other moons of Saturn are quite dwarfed by the single
giant, Titan, 3,500 miles in diameter. Because of its methane
atmosphere, this world may one day play a very important part
in the exploration of the outer planets. Methane (CH_4) appears,
as far as we can judge today, to be an excellent propellant for

Plate IX

THE MARTIAN BASE

Painting by Leslie Carr, based on a drawing by R. A. Smith

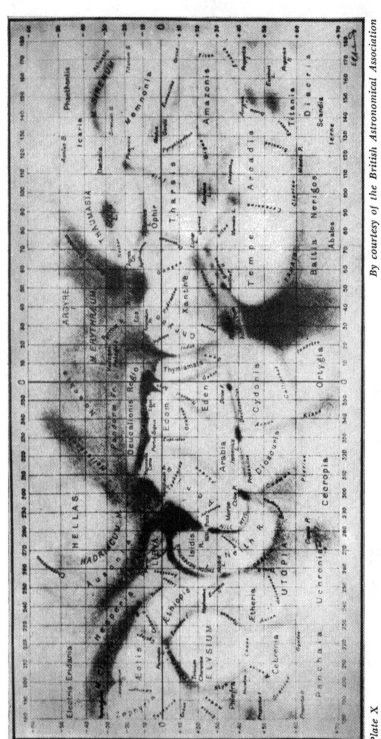

Plate X

MAP OF MARS

By courtesy of the British Astronomical Association

atomic rockets, so it is quite conceivable that this world may be an invaluable refuelling point!

Although it does not require much more energy to reach the orbits of the outer planets than it does to reach Mars or Venus, the journeys last very much longer if the "minimum energy" routes are followed. (Figure 10.) The one-way journey to Jupiter, for example, takes 2 years 9 months, and the complete round trip would last at least six years. For the planets beyond Jupiter, the times would be still longer. However, if refuelling were possible much more direct routes at far higher speeds would become practicable.

Saturn's most characteristic feature is, of course, its wonderful system of rings, which, even in a small telescope, presents a spectacle so extraordinary that one never quite gets used to it. The rings consist of myriads of particles (some of them probably little larger than dust) travelling round the planet in absolutely circular orbits, all lying in the same plane. Saturn's rings are, indeed, the nearest thing to a geometrically perfect plane that Nature has ever produced. When seen edge on, as happens at intervals of about fifteen years, the rings vanish completely.

From its inner moons, Saturn must present a glorious sight. Thus in the sky of Mimas, the closest satellite, it would fill almost 5,000 times as much of the sky as does our own Moon. It would also go through its phases much more rapidly, since Mimas completes a circuit of Saturn in less than a day. The cloud belts and occasional great atmospheric disturbances could be observed in minute detail and, as Mimas cannot have much weather of its own, would provide any inhabitants with an alternative and inexhaustible topic of conversation!

There appears to be something very peculiar about the smaller of Saturn's moons, for their densities (particularly in the case of Mimas) work out at much less than that of water. This might mean that they are highly porous (perhaps, as Hoyle has suggested, "gigantic snowballs"). It is equally possible that our measurements are incorrect, for it is extremely difficult to determine the size of such small and distant objects with any degree of precision.

We know very little about Uranus and Neptune because their immense distances make it impossible to observe them success-fully except in the very largest telescopes. Once again they are giants with tremendously deep methane atmospheres, but pos-sibly because of their extreme coldness they do not show the disturbances which can be seen on Jupiter and, to a lesser extent, on Saturn.

Uranus has five satellites—Ariel, Umbriel, Titania, Oberon, and the recently discovered Miranda. (Presumably the next to be detected is doomed to be christened Caliban.) Titania is about half the size of our Moon, but the others are much smaller.

Neptune has only two moons, Triton and Nereid. Triton is one of the largest known satellites, being about 3,000 miles in diameter, but for some reason it has never been officially named and the "Nautical Almanac", the arbiter in such matters, always refers to it long-windedly as "the satellite of Neptune". Perhaps the discovery of little Nereid in 1949 may expedite the formal christening of its big brother.

Until 1930, Neptune's orbit marked the frontier of the Solar System. In that year Pluto was discovered as a result of a long search by the Lowell Observatory. The discovery was based on mathematical calculations by Dr. Lowell, but it has now been found that Pluto cannot be the planet whose existence he pre-dicted! It is far too small—having a diameter of less than 4,000 miles—so we can only assume that its discovery was fortuitous and that the planet for which Lowell was looking still remains to be found. Although nothing is known about Pluto except its size and its orbit, it probably resembles the inner planets in composition and so will have nothing in common with its giant neighbours. It must be exceedingly cold, the temperature never rising above − 350 degrees F. Almost all gases except hydrogen and helium would be liquefied at this temperature, so it is not likely that Pluto has an atmosphere.

This exhausts the list of planets, as far as it is known today. There is no theoretical reason why planets should not exist at very much greater distances than Pluto, but their detection would be extremely difficult and to some extent a matter of luck.

That the Sun can control the movement of bodies far out into interstellar space is proved by the existence of comets. Some of these strange entities travel on extremely eccentric orbits that take them hundreds of times further from the Sun than Neptune or Pluto. Their origin and structure involve problems which are still unsolved, and as they are almost entirely composed of gas at a density which we would consider a good vacuum they are not of much interest to astronautics. Some of the larger specimens may have a solid core but this probably consists of nothing more than a cluster of meteors.

Planets—satellites—asteroids—comets—this completes our survey of the Solar System. In Chapter 16 we will discuss the probable existence of other planetary systems around other stars, but as far as our present definite knowledge goes, the only possible abodes of life in the universe are the worlds we have been describing. Most people will probably feel that the resulting picture is not exactly an encouraging one. They may be right: there may well be no advanced form of life, in our Solar System, beyond the atmosphere of the Earth, and no life of any kind except a few lichens on the Moon and Mars. Yet there is a danger that this assumption, plausible though it may seem, is based on a hopelessly anthropomorphic viewpoint. We consider that our planet is "normal" simply because we are used to it, and judge all other worlds accordingly. Yet it is *we* who are the freaks, living as we do in the narrow zone around the Sun where it is not too hot for water to boil, and not too cold for it to be permanently frozen. The "normal" worlds, if one takes the detached viewpoint of statistics, are the Jupiter-type planets with their methane and ammonia atmospheres.

We do not know the limits to the adaptability of life. On our planet, it has learned to function over a temperature range which is equivalent to a move from Venus to Mars. It is based on oxygen, carbon and water, which are among the most abundant substances in the crust of the planet. Yet these basic materials are utilised in very varied fashions. Some organisms (e.g. jellyfish) consist almost wholly of water: others, such as cacti, use very little and survive in environments too dry for any other form of life. Certain bacteria have even performed the astonishing feat of partly replacing carbon by sulphur, and can live happily in boiling sulphuric acid.

The importance of water arises from the fact that it dissolves such an enormous variety of substances and so acts as a medium in which countless chemical reactions can take place. In this respect, however, it has a number of rivals—liquid ammonia among them. On a planet whose temperature was less than —28 degrees F. but above —108 degrees F., ammonia might take the place of water for many purposes. On even colder worlds methane, which remains liquid down to the extraordinarily low temperature of —300 degrees F., might take over. It is true that most chemical reactions proceed very slowly, if at all, at low temperatures. However, fluorine, the most reactive of all elements, could conceivably replace oxygen under these conditions.

In the direction of increasing temperatures, it is again difficult to set a limit to Nature's ingenuity. The discovery of silicon-carbon compounds in the last decade has opened up new vistas in organic chemistry, and a life-form based partly on silicon is by no means beyond the bounds of possibility. The silicon compounds retain their identity at temperatures high enough to destroy their carbon analogues and they might make life possible on worlds a few hundred degrees hotter than Earth—for example, on parts of Mercury.

Faced with an unpromising environment, life has the choice of two alternatives—adaptation, or insulation. Examples of both can be seen on our world. In the polar regions, the seals and penguins adapt: the Esquimos insulate. One of the most remarkable examples of this last technique is provided by the humble water-spider, a wholly air-breathing insect which nevertheless spends much of its time submerged. By carrying its appropriate living conditions with it, it manages to survive in a completely alien environment. In the same manner, carbon life based upon water could conceivably exist even on the frozen outer worlds. One can imagine beings with tough, insulating skins through which the heat loss would be very small. As long as they had some source of energy—chemical, solar, perhaps even nuclear—and the necessary food, they could still survive though their surroundings were not far above absolute zero.

It may be objected that though such life-forms might be able to exist on very cold worlds, they could hardly have *originated* there. The indigenous life would probably be based on low-temperature reactions and would not be much hotter than the surroundings. Yet from this type of organism higher forms of life might be able to evolve, just as the warm-blooded mammals evolved from the cold-blooded reptiles.

We know, of course, practically nothing about the laws which govern the appearance and the evolution of life on any planet. The above speculations may help to show the danger of generalising from the solitary example of our own Earth, and trying to produce laws applicable to totally alien planets. It is illogical to be depressed because the other worlds of the Sun are so different from our own that we cannot hope to find familiar forms of life there. These very differences will make their exploration all the more interesting. After all, interplanetary travel would lose much of its point if the other planets were simply new editions of Earth! It is well to remember that there are only three worlds in the Universe whose surfaces we can observe at all closely. They are Earth, Mars and the Moon. On these, as far as we can tell, life has scored one hit, one probable, and one possible. Many astronomers would go so far as to say two hits and one probable. That is not a bad beginning, and it leads us to hope that the average may be maintained elsewhere.

14. Exploring the Planets

I was thinking this globe enough till there sprang out so
noiseless around me myriads of other globes.

WALT WHITMAN—*Night on the Praries*

BY now the reader may have begun
to suspect—and rightly—that the problems of space-flight, difficult
as they may be, will be quite surpassed by some of those we shall
have to face when we begin the actual exploration of the planets.
This is a point which is often ignored, or hastily by-passed, by
writers on astronautics, who no doubt feel that they already have
enough to deal with. Yet if our survey of the subject is to have
claim to thoroughness, we cannot overlook these difficulties.

The speed with which the planets are explored and all their
possibilities developed depends upon factors many of which are
outside the purely technical domain. Although the physical conditions encountered, the supplies of minerals located, the nature
of the atmosphere, and so on, will clearly be of fundamental
importance in determining what *could* be done upon any planet,
what *will* be done is likely to be decided by the resources made
available from Earth. This takes us from the realm of science into
that of economics and politics, and raises questions which we
will defer until Chapter 18. In what follows now we will assume,
perhaps optimistically, that one day the human race will be
prepared to devote to astronautics the effort and money needed
to wage a minor war.

The exploration and colonisation of the Moon has already
been discussed in Chapter 11, and much that has been said there
can be applied, with relatively little change, to Mars, Venus and
Mercury. In many respects Mars and Venus will present simpler
problems than the Moon, for the presence of an atmosphere—
unless it is violently poisonous or corrosive—is always an advan-

tage, even if it cannot be breathed. It means, in particular, that the wearing of bulky space-suits, and the complete pressurisation of all buildings, is unnecessary.

Although we do not know the composition and pressure of the Venusian atmosphere at the surface of the planet, it is certainly at least as dense as Earth's, but probably not so compressed that it would be dangerous to human life. Thus while we may have to carry oxygen on Venus, we should require no additional mechanical aids—there will be no need for the radio sets, the refrigeration and the heat insulation that was essential on the Moon. This at least should be true for some parts of Venus, though it may not be so for all.

The presence of an atmosphere also means that, for periods of several minutes, men could go out into the open with no breathing equipment at all. This is one of those practical points which, though not very important in itself, would make life on Venus much simpler than life on the Moon. It is a very surprising fact—and one which is not at all well known—that after inhaling pure oxygen for some time one can then carry on *without further breathing* for periods of up to ten minutes. (The record is actually 15 minutes!)

An atmosphere of reasonable density would also simplify the construction of buildings, because it could be arranged that the pressures inside and out were the same, even if the composition of the gases were quite different. This would mean that the danger of leaks or explosive decompression, always present to some extent on a world like the Moon, would be removed.

The existence of an atmosphere also means that air transport would be possible, with all that this implies in ease of travel and exploration. As we have pointed out in Chapter 11, rockets do not appear to be practical, and would certainly be very uneconomic, for journeys of a few hundred miles, so that on the Moon all short-range transport would have to be confined to the surface. We would no longer be faced with this restriction on Mars and Venus, and air-borne flying machines would be feasible.

Of course, our present-day internal-combustion engines would be useless in the inert atmospheres of these planets, but as we are looking at least fifty years ahead it is reasonable to suppose that by then nuclear power plants for aircraft will have been perfected, and the surrounding atmosphere will be used merely as the "working fluid", taking no part in any chemical reaction. (And playing, in fact, the same rôle in aviation that water does in ocean transport.)

Until we have more definite information about the density of the Martian atmosphere, we cannot be sure if men will be able to live there without pressure suits and wearing only breathing masks. At the moment, the evidence is certainly against this, but the problem is a "marginal" one.

There is little doubt that aircraft could function in the Martian atmosphere, despite its thinness. The fact that gravity has only a third of its terrestrial value would be a great help to flying, and the aircraft employed would probably be high-speed machines with low wing-loadings. They would keep relatively near the surface, to remain at the level of greatest air density, but on a flat planet like Mars this would not involve any danger.

Since the astronomical evidence suggests that oxygen may be present in very large quantities on Mars and Venus, though in each case combined with other elements, the provision of a breathable atmosphere is a problem of chemical engineering which is, in principle, capable of a solution. As we have already remarked, there is no limit to what may be done if one has power, knowledge, and the necessary raw materials. On both Mars and Venus it would seem possible to build very large pressurised domes, big enough to enclose whole settlements or even small towns, yet using no such supporting members as arches or columns. They would be held up by air pressure alone—a method of support which, surprising though it seems, would allow the roofing of very large spaces on Earth and would be still more promising on planets of lower gravity.

A horizontal sheet with one atmosphere pressure beneath it

and zero pressure above would have a vertical force of nearly a ton acting upwards on every square foot. A light fabric or balloon structure can therefore be held to a rigid shape, and will carry considerable loads, if the pressure excess between inside and out is even a very small fraction of an atmosphere.

Probably the largest gas-supported structure yet built was the envelope of the stratosphere balloon *Explorer II*, which formed a sphere nearly 200 feet in diameter when fully inflated. On a planet like Mars, there would be no fundamental reason why hemispherical domes a thousand feet or more in diameter might not be constructed. Inside these great bubbles of air the colonists could live exactly as they would on Earth: only when they ventured outside would they have to put on their breathing equipment. If desired, the domes might be made of some transparent, flexible plastic to let through the sunlight, though this is by no means essential and might result in too great a heat loss during the night. The best arrangement would be a dome which was transparent during the day, and so collected heat on the "greenhouse" principle, and which could be made opaque at night.

An impression of such a Martian colony is given in Plate IX. On the excellent principle of not putting all one's eggs in the same basket, there would be several small pressure domes rather than a single big one. They would be linked together with air-locks and there would also be locks in each dome communicating with the surrounding countryside.

In Plate IX, the residential and administrative area is in the foreground. Although as the "weather" inside the domes would be completely under control, it would appear advisable to roof the buildings so that they could be individually pressurised. Thus in case of failure of the dome all the inhabitants indoors would be safe, and could go to the help of anyone who had been caught out in the open. (It should be realised that such a failure would never be instantaneous and there would normally be time for anyone out of doors to reach safety.)

The further domes cover the chemical plant—where oxygen and other essential materials are obtained from ores and minerals brought in from outside—and the food-producing plant with its farm and processing equipment.

In the distance is the combined air- and space-port. Here the atmospheric-type rockets land after making contact with orbiting spaceships: it is quite possible that one of the tiny moons— either Phobos or Deimos—would be used as a rendezvous for this purpose. A jet aircraft is seen departing on a journey to another settlement. For short-range transport, pressurised vehicles with large balloon tyres would probably be used, as on the Moon.

Within such cities, the lives of the Martian colonists need not be unduly restricted or monotonous. Boredom would, in any case, be the least of their worries. Around them would be a whole world awaiting discovery—a world which will probably keep geologists, botanists and zoologists busy for centuries.

It is difficult to say how long it would take to build bases on this scale, for we know too little about conditions upon Mars. When we reach the planet, we may discover factors which change the situation radically—for better or for worse. Plate IX must not, therefore, be taken too literally. It is merely an attempt to show how we could deal with one particular environment: the actuality will certainly differ in important details.

It should also be unnecessary to point out that everything shown in Plate IX would have been made from local materials. Only the essential tools would have come from Earth originally, and every effort would be made to put the colony on a self-sufficient basis at the earliest possible date.

The remoter planets and satellites present much more severe problems owing to their extreme coldness. These problems, however, should not be exaggerated. The heat loss through properly designed multiple-walled structures can be made very small and as far as permanent buildings are concerned it is not much more difficult to insulate against —400 degrees F. than —100 degrees F.

Care has to be taken in the choice of materials, many of which change their properties entirely at these low temperatures. In general there is an increase of strength accompanied by greater brittleness. There would, of course, be no difficulty in providing all the heat needed to keep an encampment warm even on Pluto. Heat is the one commodity which atomic power can already supply in unlimited amounts.

It would be when one attempted to explore these planets, either in space-suits or vehicles, that the cold would prove a difficulty and a danger. Heat can be lost from a body in three ways—by conduction, radiation and convection. On an airless planet, the latter effect would not exist and the only conduction loss would be at the points of contact with the ground. Space-suits would therefore have to be fitted with thick insulated boots, perhaps with corrugated soles to reduce the area of contact. In this way the loss of heat by conduction could be made very small.

Radiation loss would be reduced by use of the "vacuum flask" principle—that is, by employing double walls suitably silvered to reflect heat back into the interior. It is obviously easier to apply this construction to vehicles than to space-suits.

Conditions on the other planets will probably stimulate the development of robots which can replace man in hazardous situations. Most of the technical problems involved in the construction of such machines have already been solved. Anyone who has ever watched the mechanical hands and arms at work behind the lead barriers in a laboratory dealing with radioactive isotopes will agree that the remotely controlled robot is already almost with us.

When one considers some of the heavy machines which we have developed during the last generation to change the face of our own planet—the tractors, bulldozers, graders, excavators and so on—it requires little extra imagination to picture more advanced machines, suitably modified, working for us on other worlds. They would have the advantage (shared by the men who controlled them) of operating under lower gravity than on Earth.

Their greatest enemy would be cold, which if not guarded against would cause sudden fractures through crystallisation of the metal.

It is worth remarking that, on any planet, the problem of conserving heat would be greatly simplified by going underground, as rock is an excellent insulator. Powdered rock, with air or vacuum between the particles, is the nearest thing to a perfect non-conductor of heat that is known.

When we possess spaceships which can reach the orbits of Jupiter and Saturn in a reasonable length of time, attempts will no doubt be made to set up small scientific observatories, if nothing more, on some of the many moons of these giant worlds. As we have very little idea of what we may find on any of these satellites (several of which, it should be recalled, are fair-sized planets themselves) we cannot guess what efforts it may be worth expending in this direction.

The range of possibilities is enormous: worlds like Titan, Ganymede, Callisto and Europa may be barren lumps of ice, or they may teem with extraordinary life-forms perfectly adapted to low-temperature existence. They may be of no interest geologically, or they may be rich with minerals unlike any found on Earth. Just where the truth lies between these extremes, we shall not discover until we have gone there to look for ourselves.

At some unknown distance in time after the establishment of enclosed colonies on the planets, attempts will almost certainly be made to make much larger areas suitable for human life. The provision of an atmosphere for an entire world cannot be regarded as a technical impossibility, and even altering the temperature of a planet by several hundred degrees would not be out of the question to the scientists of an age which had completely mastered nuclear energy.

Exploring this theme, Professor Fritz Zwicky, who is not only one of the world's leading astronomers but is also research director of a very large American rocket engineering organisation, has suggested that eventually we may learn how to alter the *orbits* of the other planets so that their climates may be controlled. Fantastic though this proposal may seem, it does not involve any inherent impossibility, and who today would dare to put any limits to the science of five hundred or a thousand years hence?

15. Stations in Space

Up above the world you fly,
Like a tea-tray in the sky.
LEWIS CARROLL—*Alice in Wonderland*

WE now come to a subject which, chronologically, might well have been considered earlier in this book, as its achievement will certainly take place sooner than the long-term projects we discussed in the last chapter. Indeed, many believe that the building of the space-station may be the first task of astronautics, antedating even the journey to the Moon. If one uses the word "space-station" to describe any artificial structure in a permanent, stable orbit, this view is certainly correct, for there is no doubt that instrument-carrying missiles will be established beyond the atmosphere at a fairly early date. Piloted missiles, remaining in their orbits for relatively short periods, will follow soon after. It is better, however, to restrict the term to permanent manned bases or observatories, constructed in space by materials ferried up by rocket and kept supplied with stores and personnel by the same means.

The idea of the space-station originally arose from the conception of orbital refuelling described in Chapter 5. When it was realised that permanent structures could be established in space, it was quickly seen that they would be useful for so many scientific purposes that their employment as "filling stations" for rockets might well become of secondary importance. Indeed, there is probably no need to use space-stations at all for this purpose, at least in the earlier stages of interplanetary flight, since the first spaceships will be refuelled directly from other rockets, as shown in Plate II. Not until there are large numbers of ships coming and going at frequent intervals would it be worth while setting up stations exclusively for refuelling purposes.

Let us briefly recall the underlying principles involved in the creation of any form of artificial satellite. It will be remembered that, just outside the atmosphere, a body travelling horizontally at 18,000 m.p.h. would remain perpetually in a stable, circular orbit, without requiring any power, and making a complete rotation round the Earth in a little over ninety minutes. At greater heights, the orbital speed needed is less and so the period of revolution increases; 22,000 miles above the surface the period is exactly 24 hours, and a body here, if originally above the Equator, would revolve with the Earth so that it would neither rise nor set. An object at a greater distance would move more slowly than the Earth on its axis and so would rise in the east and set in the west, as do all the celestial bodies. Inside this limit of 22,000 miles a satellite would rise in the *west* and set in the *east*, as does the inner moon of Mars. (Anyone who cares to deduce from this that the Martians are builders of space-stations is welcome to do so!)

Bodies orbiting the Earth need not, however, travel on circular paths: any ellipse with the centre of the Earth at one focus is a possible orbit—so long, of course, as it does not intercept the atmosphere. Nor need the orbits lie in the plane of the Earth's own rotation: they could be at any angle to the Equator, and could for example pass over the Poles. Which orbit was selected would depend on the purpose which the satellite was intended to fulfil.

A satellite a few hundred miles above the Equator would, because of the Earth's curvature, be visible only in a rather narrow band around the planet: and conversely, it would be able to survey only this restricted region. If its orbit passed over the Poles, however, the rotation of the planet would ensure that in quite a few revolutions of the satellite (i.e. in under a day) the whole surface of the Earth could be surveyed.

The same result would be obtained by using an orbit inclined to the Equator at an angle of, say, 45 degrees, as long as the satellite was more than a thousand miles above the surface. Only the region immediately around the Poles would remain invisible from such a satellite in the course of 24 hours—although the distortion caused by the Earth's curvature, and the thickening atmospheric haze, would make useful observation impossible well before the station's visible "horizon" was reached.

Although, as is invariably the case, we shall not discover the full value of space-stations until we have actually constructed them, some of their uses are already obvious. The most important may be listed as follows: (1) astronomical and physical research; (2) meteorology and surveying; (3) biological studies; (4) refuelling; (5) radio relaying; and (6) dock facilities. Some of these functions could be carried out on the same station, but others demand such varying types of orbit that specialised stations would eventually have to be built, devoted to a single purpose. Thus the refuelling stations would be as near the Earth as possible (perhaps only five hundred miles up) whereas the astronomical ones would be at ten or a hundred times this distance. In this connexion, however, it should be pointed out that it is much "cheaper", in terms of energy, to establish a close satellite than a distant one.

The advantages of a lunar observatory have already been discussed in Chapter 11. These would apply still more strongly to an observatory in space, which would be able to survey the complete sphere of the sky. Even the Moon's extremely tenuous atmosphere might affect certain very delicate observations: this factor would not arise at all on the space-station.

Perhaps the most interesting possibility opened up by the space-station—although this is admittedly looking a long way ahead—is the fact that it would remove the restrictions imposed by gravity on the size of astronomical instruments. The great difficulties involved in building the 200-inch telescope on Mount Palomar were not primarily optical ones: they were largely due to the fact that the mirror and its auxiliaries had to remain rigid to a few millionths of an inch, no matter how the enormously heavy instrument was tilted and revolved.

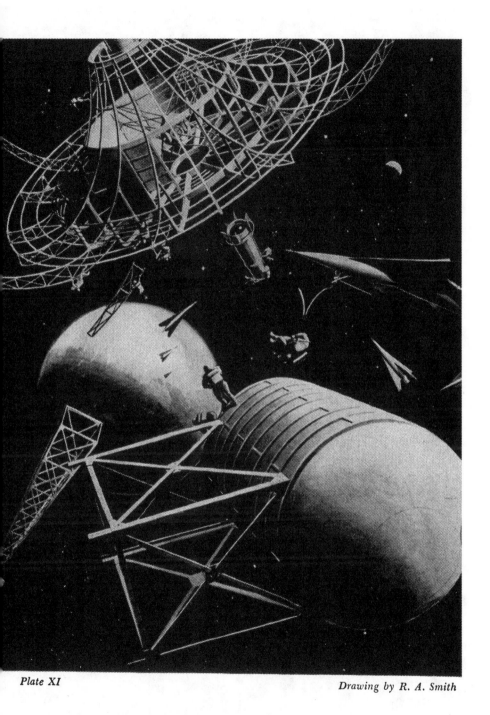

Plate XI   *Drawing by R. A. Smith*

BUILDING THE SPACE-STATION

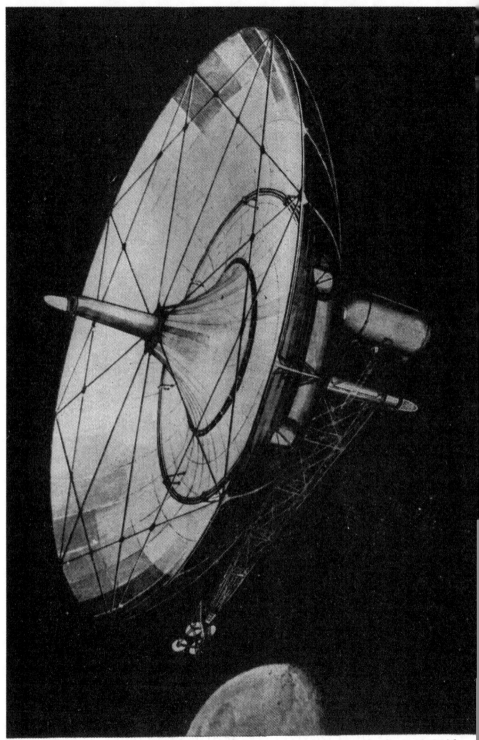

Drawing by R. A. Smith

THE SPACE-STATION

In the weightless condition which always applies to a body in free orbit, the telescope structure need have only enough strength to maintain its stiffness. Indeed, the optical elements might be miles apart, if need be, with no physical connexion at all. This would make it possible, for the first time, to build instruments which could measure the diameters of normal-sized stars. It might even become possible to detect planets of the nearer stars, something quite out of the question with earth-based equipment.

Since 1945, astronomers have become more and more interested in "telescopes" employing not light rays but radio waves, which have much greater penetrating power and so may be able to teach us something about the structure of parts of the Universe which we can never "see" in the ordinary way. Because radio waves are about a million times longer than light waves, it is necessary to build instruments of enormous size to get reasonable definition. The biggest radio-telescope yet built is over 200 feet in diameter (compared with 200 *inches* for the largest optical telescope) but even so its resolving power—that is, its ability to separate close objects—is thousands of times poorer than that of a pair of cheap opera-glasses. Moreover, because of its size, it is incapable of being moved.

Out in space, these limitations could be overcome, for it would be possible to build radio-telescopes literally miles in diameter—and still, thanks to their weightlessness, make them movable.

The assembly and operation of giant telescopes (optical or radio) floating in space clearly involves engineering problems of no mean order. Some of these will be considered on page 159, when we will discuss the actual construction of space-stations.

The opportunities which an artificial satellite would provide for physical research are equally great. A vast new field of experimental science would be opened up by the condition of weightlessness, and the presence of a virtually perfect vacuum of unlimited extent would be a stimulus to such studies as electronics, nuclear physics and the innumerable branches of technology which demand the use of low pressures. It would also be possible, for the first time, to produce temperatures not far from absolute zero over large volumes of space.

The study of cosmic rays (one of the key problems of modern physics) would receive an immense impetus, since only outside the atmosphere can we observe the primary radiations. Our knowledge of the ionosphere, which is of great practical impor-tance in radio communication, would also advance rapidly once we were able to observe it from both sides.

Since a space-station a few thousand miles up would be able to survey the greater part of the planet in a couple of hours, watching all cloud formations and the movement of storm centres, it could clearly play a very valuable rôle in meteorology. Although ground stations would presumably still be necessary to fill in details of atmospheric pressure, temperature and so on, the satellite would be able to give the overall picture almost literally at a glance.

An orbital satellite (even one carrying no men or instruments) would also be invaluable as aid to navigation, provided that it was bright enough to be observed visually from the Earth. For centuries navigators have found their way over the surface of this planet by the use of sun, moon and stars. The existing heavenly bodies, however, are too far away for certain simple and direct position-finding methods to be employed, and one or two close satellites would greatly ease the problem.

The use of space-stations for biological research is a somewhat more speculative matter, for no one can say what medical science may discover from the study of organisms living for long periods under zero gravity. It is worth remembering that gravity is an important factor determining the possible size of micro-organisms (and indeed larger creatures). Its removal might produce some interesting results, though whether we could breed amœbæ as big as footballs remains to be seen!

The absence of gravity would certainly give medicine a most

important new weapon, and not only for the treatment of obvious complaints like heart disease. It would probably accelerate any form of convalescence,.and it is not too fantastic to suggest that many of the hospitals of the future will be found in space. We have seen that even the use of chemically fuelled rockets need not subject passengers to abnormally high or dangerous accelerations, so there is no reason why invalids should not travel in spaceships almost as safely as could people in normal health. (It is amusing to think that thirty years ago the idea of transporting the sick by air would have seemed complete madness—yet nowadays it is often the preferred method!) The fact that, apart from relatively short periods of eclipse by the Earth, a space-station would be in continuous sunlight would be of great therapeutic value. So would be the spectacle of the Earth itself, almost filling the sky and going through its phases from new to full in a few hours. The infinite variety of detail presented by the continents, seas and clouds, the pleasure of picking out familiar landmarks and even of observing the streets of the great cities through telescopes, should reconcile the patients to their temporary exile. The "Earthside" ward of a space-station hospital would, indeed, be a room with a view!

The refuelling and repair bases would probably be in the closest and hence most economical orbits. They might eventually become very extensive affairs—real "space-ports" with elaborate harbour facilities and huge hangars which could be pressurised to assist repair work. One would expect to find in them everything which a present-day seaport or airway terminus can provide.

Even when it becomes technically possible to make journeys to the other planets direct from the surface of the Earth, it seems doubtful if spaceships will in fact do so. The advantages of orbital techniques are so overwhelming that the time may never come (at least while the rocket is the only means of propulsion available) when interplanetary ships will take-off from the Earth. The orbital space-ports may therefore expect a long period of useful service before they become technologically obsolete.

On a purely commercial basis, the greatest value of a space-station would probably arise from the radio and television services it could provide. An orbital satellite would make possible, for the first time, a reliable system of radio communication between all points on the Earth, irrespective of ionospheric conditions, magnetic storms, and all the other vicissitudes which plague long-distance radio. It would also vastly improve the quality of the service, for there would no longer be any need to rely on reflected waves with their subsequent distortions and interference. A direct beam service could link the station with any point on the hemisphere below, and messages could be relayed on to any other point around the curve of the planet—if necessary through a second space-station. Complete coverage of the whole Earth would be provided by three stations, revolving in the same orbit but spaced 120 degrees apart. (Figure 18.)

The obvious orbit for this purpose would be the 24-hour one, 22,000 miles above the Equator. Any point on the Earth's surface would then have at least one station permanently visible in the sky—and, moreover, *fixed* in the sky, unlike those wayward bodies the Sun, Moon and stars.

Perhaps the most exciting prospect raised by the relay chain is that it would make a world-wide system of television practicable. It is, indeed, almost impossible to imagine any other way in which this could be done, since the curve of the Earth limits the range of all surface transmitters, no matter how powerful, to less than a hundred miles. Three stations in space, linked to each other by microwave beams, could provide a television service *over the whole planet* for no more power than one of today's larger transmitters.

We have heard it said that it would be difficult to think of a better argument against space-travel than this. However, even if one takes a pessimistic view of television's cultural future, it should be pointed out that high-frequency waves of the type it uses have many other duties. They would make possible an almost unlimited number of interference-free communication

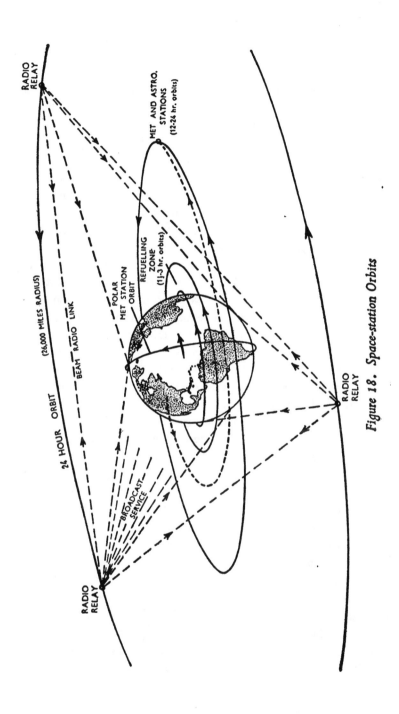

RADIO RELAY

MET AND ASTRO. STATIONS
(12-24 hr. orbits)

REFUELLING ZONE
(1½-3 hr. orbits)

POLAR MET STATION ORBIT

(26,000 MILES RADIUS)

BEAM RADIO LINK

24 HOUR ORBIT

BROADCAST SERVICE

RADIO RELAY

RADIO RELAY

Figure 18. Space-station Orbits

channels, and would provide navigation and air safety services beyond anything in prospect today. A world society must possess a fast and reliable system of communications. The use of radio relays in space could provide this on a scale quite impossible by any other means, and at very great economy.

Shortly after the end of the Second World War, a good deal was heard in the Press about gigantic "space-mirrors" which the Germans were supposed to be considering as weapons of war. The truth behind this rumour was somewhat less spectacular, as is usually the case. Twenty years earlier, Hermann Oberth had pointed out that it would be mechanically possible to build enormous mirrors in space by assembling sheets of metallic sodium on a spider's web of cables, kept rigid by its own rotation. A mirror two miles in diameter would collect about 10,000,000 h.p. of solar energy and if this were directed on the Earth beneath it could produce local heating over a fairly large area. Oberth imagined the use of such mirrors for weather control by the evaporation of water in some areas and the "directing" of the resultant vapour to others. The laws of optics would make it impossible to focus all the heat from such a mirror so accurately that it could be a really dangerous weapon, though if the mirror were made a good deal larger it might be able to make selected areas uncomfortably warm.

In any case, as a weapon the space-mirror must be regarded as obsolete before it has been built. We now know much more effective methods of incinerating populations, and the mirror itself would be an exceedingly vulnerable structure—it would take years to build yet could be destroyed in a few minutes by half a dozen guided missiles.

In Figure 18 the orbits for the various kinds of service discussed above are shown. Although all the orbits hitherto mentioned have been circular, highly elliptical ones might be utilised for special purposes. These would take the station perhaps hundreds of thousands of miles away from the Earth and bring it back, a week or so later, to within a few hundred miles of the atmosphere.

It might be thought that the various orbits shown in Figure 18 would interfere with each other, and that there would be some danger of collisions. This, however, is quite impossible for the stations in circular orbits, which could not change their distances from the Earth without the use of power. Spaceships pulling away from the refuelling zone would, of course, have to time their departure to avoid passing too close to an outer station. Since there are so many billions of cubic miles involved, this would not be a serious restriction on flight orbits—which in any case would usually depart from the equatorial plane at quite short distances from Earth.

At least one complete book has already been written about the construction, planning and internal mechanics of space-stations. Here we can mention only a few of the more important aspects of this intriguing subject, which will certainly present the engineers of the future with some very peculiar problems.

The first stations will probably be built from spaceships that have been "cannibalised", to use a horrid but expressive word, once they have reached orbital velocity. The assembling of the structure will be done by men in space-suits, propelling them-selves and their loads by reaction pistols or gas jets. Perhaps instead of what has now become the "conventional" space-suit, tiny one-man spaceships might be used, just large enough to hold a single occupant and fitted with the necessary handling mechanisms.

Relays of rockets would climb up into the orbit with loads of stores and structural materials and would simply "dump" their cargoes in space until they were needed. The assembling would of course be enormously simplified by the absence of gravity, and the first structure to be built would probably be a spherical chamber which could be pressurised to serve as living quarters for the space-station staff. From this beginning, endless variations of design are possible. Some stations might be in the shape of flat discs, slowly spinning so that there would appear to be normal gravity at the rim. At the axis there would be no gravity at all, so in the one station it would be possible for the staff to live normal lives yet carry out zero-gravity experiments whenever they wished. Other satellites—particularly those used for astronomical research, where instruments have to be kept rigidly fixed for long periods in an unvarying direction—would have no rotation. The crews might live when off duty in an associated spinning station a few miles away, transferring from one to the other by low-powered rockets.

Over the years and centuries the stations would grow by a process of accretion as new chambers and laboratories were built, until they might eventually become miles in extent, forming a loosely connected group of structures serving many different purposes—in fact, veritable cities in space. The larger ones would be easily visible from the Earth as bright stars crossing the sky rapidly from west to east, but often vanishing for a while as they passed into the shadow of the planet and were "eclipsed".

The construction of one possible type of station is shown in Plate XI. This particular model is designed to have centrifugal gravity and is fitted with a solar power-plant. (The parabolic girders will eventually be covered with sheets of reflecting material to concentrate sunlight on the boilers.) In the foreground is the pressure-chamber through which access to the interior of the station can be gained, while floating in the middle distance is a lattice mast which will be used to support radio equipment.[1] The complete station is shown in Plate XII.

Although it has been assumed that, once set revolving in its orbit, an artificial satellite would continue to circle the Earth for ever with clockwork precision, this is not perfectly correct. The disturbing effect of the Sun and Moon, and the attraction of the Earth's own equatorial bulge, would slowly alter the orbit of a satellite, causing its plane of rotation to precess or tilt slowly up and down. The effect is very small and in most cases

[1] Detailed plans of the complete station, the conception of which is due to H. E. Ross and R. A. Smith, are given in *Interplanetary Flight*. (Temple Press.)

would not be of any practical importance. It would always be possible to readjust an orbit, if necessary, by the use of minute amounts of rocket power. Indeed, some care would have to be taken to prevent accidental "perturbations" of the orbit when waste material was ejected from the station or fresh stores were taken aboard.

No doubt with the further development of astronautics artificial satellites will be built elsewhere than near the Earth. Often there would be existing asteroids or small moons which might provide a foundation on which to work and could also supply much of the necessary structural material. In this case, there would be little distinction between a space-station and a planetary colony.

Although orbital bases circling the planets might be built for numerous reasons, many might travel on independent orbits around the Sun and would thus be artificial planets rather than artificial satellites. One obvious use for such bases would be in connexion with interplanetary communication. It would fairly often happen that a spaceship or a planet would be on the far side of the Sun from Earth and so out of touch by direct transmission. A space-station moving in the Earth's orbit, but some scores of millions of miles from our planet, could be used as a repeater or relay and so would enable us to "see round the Sun".

The laws of celestial mechanics show that a small body could travel in the Earth's orbit in this fashion if it formed an equilateral triangle with the Earth and the Sun. There are two positions where this is possible—one 93,000,000 miles "behind" the Earth and the other the same distance ahead of it. Similar stable positions exist in the orbits of the other planets: in the case of Jupiter they are already occupied by two groups of asteroids known as the "Trojans".

If we take the long view of humanity's future, the time may come when these artificial worlds may be as important as the original, natural planets. In his book *The World, The Flesh and The Devil* (one of the most astonishing flights of controlled scientific imagination ever made) Professor J. D. Bernal has taken this idea to what must surely be its ultimate conclusions. He imagines spherical planetoids many miles in diameter, with food-producing areas immediately beneath the transparent outer skin. Lower still would be the machines which regulated temperature and air—which controlled, in other words, the climate of the little world.

The central volume would be the living region, which, because of the absence of gravity, would be much roomier than we, with our flat, "two-dimensional" outlook, can easily imagine. As Bernal points out, "a globe interior eight miles across would contain as much effective space as a countryside one hundred and fifty miles square even if one gave a liberal allowance of air, say fifty feet above the ground."

These worlds might develop their own cultures and specialised activities, though they would be in constant touch with their neighbours and with the various planetary civilisations. It is even possible that, eventually, only a small proportion of the human race would live upon the original planets of the Sun. A thousand years from now, indeed, the Sun's family may be very much more numerous than it is today.

Artificial worlds are also of importance because—as Bernal was again the first to point out—they might provide one solution to the exceedingly difficult problem of interstellar flight, a subject which we shall discuss in Chapter 17.

16. Other Suns than Ours

Observe how system into system runs,
What other planets circle other suns,
What varied beings people every star.
ALEXANDER POPE—*An Essay on Man*

THE Solar System, with its handful of planets scattered at immense distances from the Sun, seems to consist almost entirely of empty space. Yet looked at from the cosmic viewpoint, it is a tiny, closely packed affair. Although interplanetary distances are a million-fold greater than terrestrial ones, *interstellar* distances are a million times greater still. Even light, which can pass from the Sun to Pluto in a few hours, takes over four years to reach the nearest of the stars. It is not surprising, therefore, that it was quite late in astronomical history before it was proved that the stars were actually other suns, made pinpoints of light purely because of their enormous distance.

Our Sun is a quite typical star, although it is a good deal brighter than the average. (Only three of the twenty nearest stars exceed it in brilliance, and the vast majority are far fainter.) It is one of a very large number—perhaps 100,000,000,000—of stars forming a roughly disc-shaped system known as the Galaxy. If we could see our Galaxy from outside, it would probably look not unlike Plate XIV(a), which is a photograph of the famous Andromeda Nebula.

The stars vary in size and brightness over a truly enormous range. (Here we are referring, of course, to real variations, and not to those caused merely by the effect of distance.) If, as is customary and convenient, we take our Sun as a standard, then the biggest known stars have a diameter a thousand times as great, so that they could enclose the orbits of all the planets right out to Saturn! On the other hand, the smallest stars have less than a hundredth of the Sun's diameter, being thus smaller than the Earth.

The range of luminosity among the stars is even greater. Stars 10,000 times as bright as the Sun are known—as well as stars 10,000 times fainter. With these variations of brilliance go variations of colour. Our Sun, whose light we regard as normal, is actually a somewhat yellow star. The hottest stars of all shine with a brilliant bluish-white light, and indeed the greater part of their radiation would be quite invisible to us since it would lie in the ultra-violet. In descending temperatures the colours of the stars run: white, yellow-white, yellow, orange-yellow, orange, deep orange-red. There are also suns of almost every possible intermediate colour—gold, blue, green, topaz, emerald—so that in the telescope some of the great star clusters look like collections of jewels glittering against the blackness of space.

There are eight stars (two of them double) within ten light-years of the Sun. (This unit, the distance light travels in a year, is a convenient one for measuring stellar distances. It equals 5,880,000,000,000 miles.) Only one of our nearest eight neighbours is visible, without a telescope, to us in the northern hemisphere: this is Sirius, the brightest star in the sky, about nine light-years away. (The closest of all stars is the very faint Proxima Centauri, quite invisible to the naked eye despite its distance of "only" 4.2 light-years.)

Our Sun is in a moderately well-populated region of the Galaxy, though nowhere near its centre. On a clear moonless night the sky seems full of stars more or less equally distributed around the heavens, with the pale band of the Milky Way wandering through them. This faint arch of light, which continues round the southern hemisphere and so divides the sky into almost equal parts, was a mystery to mankind until the invention of the telescope revealed that it was composed of millions of faint stars. We now know that their faintness is due only to distance, and the reason why they form a continuous band around the Earth arises purely from our location in space. When we look towards the Milky Way, we are looking along the major axis of the Galaxy so that we see the stars packed in endless ranks as far as the eye, or even the telescope, can see. When we look in other directions, however, our gaze quickly passes out through the Galaxy's relatively thin disc and we can see only a few stars—and beyond those the great emptiness in which the other island universes float. If you look at Plate XIV(a) again, you will realise that anyone living on a planet in the outskirts of the Andromeda Nebula would see a very similar band of stars around the sky.

The heart of our own Galaxy, where the stars are clustered together more thickly than they are in the neighbourhood of the Sun, lies towards the constellation Sagittarius. Plate XIV (b) gives some idea of the great star clouds in one of the denser regions of the Milky Way, which in addition to suns contains immense clouds of luminous gas—perhaps the raw material from which the stars are made.

Although the study of the stars themselves is a fascinating and never-ending pursuit, from the viewpoint of astronautics we are only interested in planets. Unfortunately, the greatest of planets would be totally invisible at a distance of a few light-years, so we have no means of telling if even Proxima Centauri has worlds revolving round it. Our views on the existence of planets in the Universe are likely, for the time being, to be determined by whether we think the Solar System to be something usual, or an astronomical freak.

Until quite recently, the latter opinion was generally held, because the only conditions under which anyone could imagine the Solar System forming seemed to demand very unusual circumstances, such as the near-collision of two stars. Today quite a different outlook prevails. We are still by no means sure *how* planets are formed, but it is felt that many, if not most, stars may possess them. Certainly among the 100,000,000,000 stars of this Galaxy alone, there must be myriads with solar systems. But which are the stars with planetary companions, and which are alone, there is as yet no way of discovering. This is a problem which may be solved when we can build observatories in space.

In one or two exceptional cases there is some evidence for bodies of planetary size revolving around other stars. The first to be discovered was in the system of the double star 61 Cygni, about 11 light-years away. This pair of faint stars has been carefully studied for over a century, and from the movements of one component the existence of a third body has been deduced. This object has about fifteen times the mass of Jupiter, or five thousand times that of Earth. It seems too small to be a sun and may therefore be a very large planet.

There is clearly no hope of detecting worlds as small as Earth by the gravitational disturbances they produce, but if we can discover even a few giant planets by this technique it is certainly a very important step forward. In particular, it refutes the old idea that there can be only one or two planetary systems in the Galaxy—for 61 Cygni is one of our closest neighbours. It would be stretching coincidence a little too far to expect a couple of solar systems within eleven light-years of each other if they were rare phenomena.

We have several times referred to "double stars" and perhaps a word of explanation is needed about these. Our own Sun is—apart from its planets—a solitary wanderer through space. Many suns, however, occur in pairs, revolving around each other under their mutual gravitation. The variety of these partnerships is immense. Sometimes the two stars are of identical types, but sometimes they are so disproportionate in size that an elephant waltzing with a gnat would not be an inaccurate comparison.

Systems of three, four, five, six and even more suns also occur, with fantastic and beautiful combinations of colour. There seems no reason why such multiple stars should not have planets, and indeed there are cogent arguments why they *should*. We do not know how double suns are formed, but whatever the process one would expect some debris to be left over and to condense into worlds. The orbits of such planets would be exceedingly complex, in some cases never repeating themselves again so that the conception of the "year" would have no meaning. The problem of contriving a calendar for such worlds would be an appalling one, but in compensation the inhabitants would have skies whose splendour we can scarcely imagine.

In Plate XIII an attempt has been made to show the sunrise on the planet of a multiple star. It should be emphasised that, fantastic though this picture may appear, the features shown in it actually exist (except the planetary landscape in the foreground, which is of course imaginary). Coming up over the horizon is a double star of the Beta Lyrae type. The two components are very large and, unlike our own sun, are not spherical. They are so close together that their gravitational fields distort them into ellipses, the longer axes pointing towards each other. Linking the two stars is a bridge of incandescent hydrogen, which sprays out from the central sun towards its smaller companion and then forms a vast, expanding spiral—a pin-wheel of crimson flame larger than our entire Solar System. It is possible that many close double stars may be surrounded by such gaseous envelopes, which would no doubt make any nearer planets rather unhealthy. But at greater distances the fire-stream would be so dispersed that it would no longer be dangerous.

Very much further away than the close double star in the foreground are the other two components of this quadruple system. The dull red giant—most of whose radiation is in the infra-red and so is invisible to the eye—is eclipsing its smaller but more brilliant companion. The eclipse is only partial, for the little white star can be seen shining *through* the tenuous outer layers of its giant companion, like the Sun through a bank of mist.[1]

---

[1] In case it is objected that since the two systems shown in Plate XIII represent extremely unusual types of star it is not likely that they would ever be found together, we would point out that almost every combination of stars is bound to occur *somewhere* in space!

It may be wondered how we can possibly know these facts, since even in the greatest telescopes all stars are merely dimensionless points of light. In the case of double or binary systems, however, we can discover the shape and size of the components by analysing the variations in brilliance as they alternately eclipse each other. The spectroscope, which enables us to measure the velocity with which the various parts are moving, fills in the remaining details.

Almost as incredible would be the view of the heavens from a planet near the heart of a globular cluster. These are great spherical swarms of stars, so closely packed towards the centre that the separate suns must be only light-hours apart, as against the light-years that normally lie between stars. There could be no such things as night and darkness on any worlds at the core of a globular cluster: the sky would be a continuous blaze of multicoloured light, in which the individual stars would be completely lost. The dwellers of such worlds would have a very limited knowledge of astronomy, for they would be unable to observe the structure of the universe through the screen of stars which hid them from the rest of space.

As will be seen from the examples in Plate XIII, stars vary greatly among themselves in physical structure, as well as in size and brilliance. Some of the giants are so rarefied that they are a million times less dense than our atmosphere: they have been picturesquely christened "red-hot vacuums"! At the other extreme are stars whose density is thousands of times greater than any substance on Earth. The best-known example of these "White Dwarfs" is the Companion of Sirius, with a density six thousand times that of lead. A matchbox of this star's material would weigh a couple of tons—but it should be pointed out that if one *did*, by some miraculous means, obtain a matchboxful it would not stay that size for even a millionth of a second. Its density is produced by the enormous temperatures and pressures inside the star, and if these were removed it would explode with a violence probably eclipsing that of an atomic bomb.

Painting by Leslie Carr

A MULTIPLE SUN SYSTEM

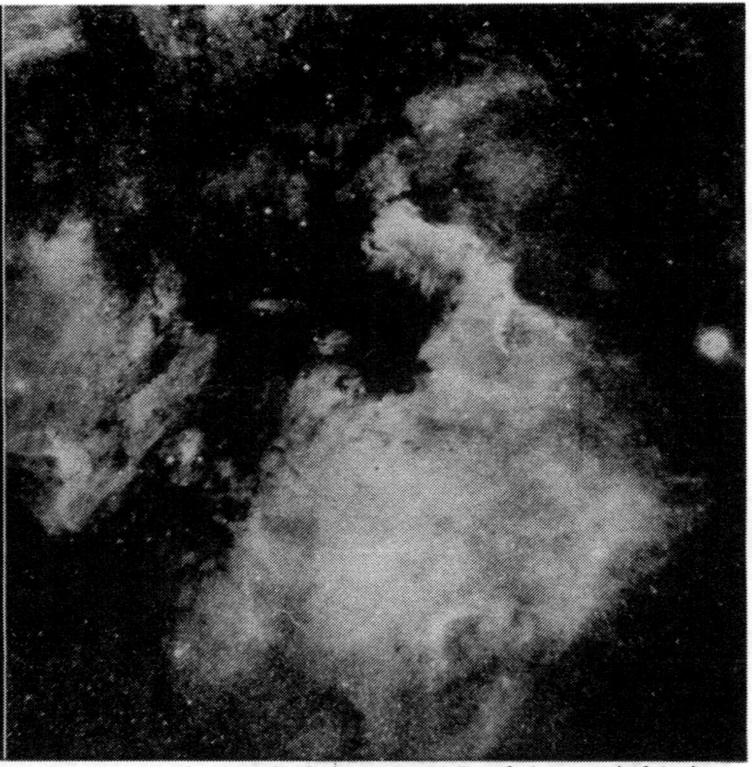

Photography by courtesy of Royal Astronomical Society

Plate XIV

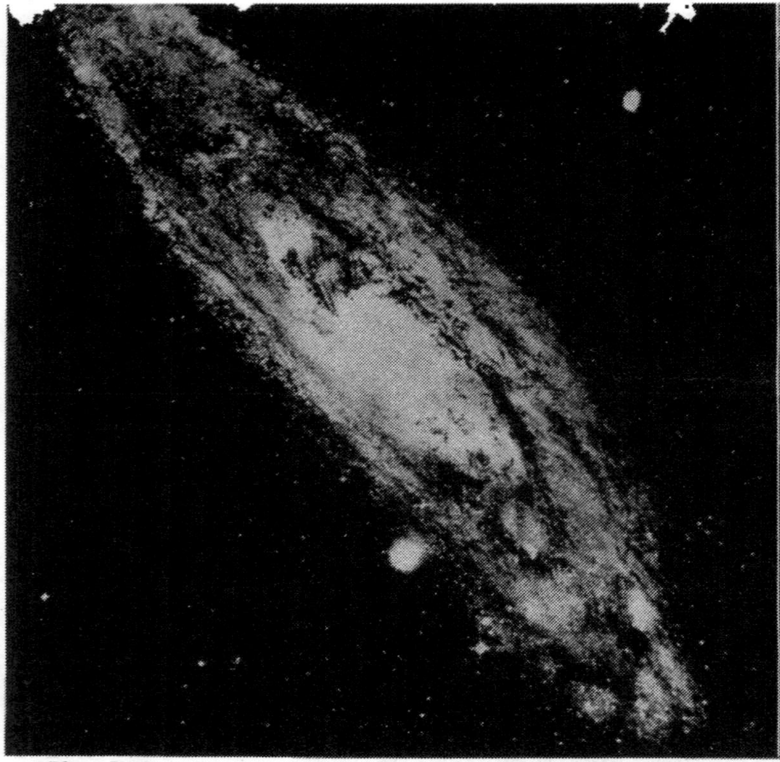

STAR CLOUDS IN CYGNUS

THE GREAT NEBULA IN ANDROMEDA

Something of this kind may indeed happen occasionally, for detonating stars ("novæ") are frequently observed. At rare intervals, they are conspicuous enough to be really prominent objects —and one (Tycho's nova, 1572) was so brilliant that for some weeks it was visible *in broad daylight.*

The cause of these gigantic stellar explosions is unknown: in their most spectacular form, a star will, within a few hours, increase its brilliance a hundred-million fold and may even, for a short while, outshine all the other suns in its universe added together. These "supernovæ" are relatively infrequent, but ordinary novæ are quite common and one cheerful theory suggests that *all* suns may become novæ at some time or other during the course of their evolution. As far as the inhabitants of any planets were concerned, the final result would be much the same whether their sun became a nova or a supernova. The difference would be, roughly speaking, that between being slowly melted or swiftly vaporised.

There are also large numbers of stars ("variables") whose brilliance fluctuates over a more modest range. Some of these stars appear to be pulsating, and they go through their cycle of brightness with clockwork precision. Others show no regularity in their variations: they behave like great, flickering bonfires, sometimes quiescent, sometimes flaring up for days or years, then relapsing again.

Such changes, as long as they were not too great, would not rule out the possibility of life on any planets of these stars. They would simply have complicated seasons—predictable in the case of the regular variables, but quite erratic for the irregulars.

It would always be possible, as far as the stars which shine with a steady light are concerned, for planets with temperatures between the boiling and melting points of water to exist. They might have to be very close to some of the cooler stars, and a long way from the brilliant blue-white suns. This would mean that their "year" might be, in the one case, only a few terrestrial days—and in the other, perhaps thousands of our years.

All the stars, including of course our Sun, are in motion through space. Their movements are not entirely random, for the great disc of the Galaxy is rotating, sweeping the stars round with it and completing one revolution in about 200,000,000 years. Since our planet was formed, therefore, the Sun has made only about a dozen circuits of the Milky Way.

This slowly turning disc of stars is about 100,000 light-years in diameter, and its greatest thickness is perhaps one-fifth of this. In the neighbourhood of the Sun (about two-thirds of the way towards the rim) the thickness of the great lens-shaped system of stars is about 10,000 light-years—though it has of course no definite boundaries.

As we look out past the thinly scattered stars away from the plane of the Milky Way, we can see, at immense distances, the other galaxies. Some we observe turned full towards us, like great catherine-wheels of stars, showing intricate and still unexplained spiral structures. Others are edge-on, still others, like the Andromeda Nebula in Plate XIV(a), tilted at an angle. In this photograph, the stars scattered over the foreground are the relatively close suns of our local system. We are looking past them, and across the immensity of intergalactic space, as a town-dweller might look past the street-lamps of his suburb to the lights of another city, many miles away.

The Andromeda Galaxy is the nearest of the other universes,[2] and it is about 700,000 light-years away. In whatever direction we look (except those in which clouds of obscuring matter block our view) we see other galaxies, extending to the limits of telescopic vision. They appear to be roughly the same size as our own system, and on the average their distances apart are of the order of a million light-years (though local clusterings occur). It will be seen that there is a breakdown here of a law which has

[2] The word "universe" is employed here in the restricted sense of a single galaxy. Thus "our universe" is merely the Milky Way system. Astronomers usually employ the word "cosmos" to describe the whole of creation—i.e. *all* the galaxies.

applied so far in the architecture of the Cosmos. The distances between stars and planets were hundreds of thousands of times greater than the dimensions of these bodies themselves. Yet the distances between the galaxies are only about ten times as great as their diameters.

The limit which we can reach with the most powerful telescope (the 200-inch reflector on Mount Palomar) is about 1,000,000,000 light-years. So far, the galaxies show no signs of thinning-out or of forming more complex structures. It is possible that we have come to the end of the hierarchy, but this is a matter concerning which we can only speculate at the moment. In the next few decades, we shall have a great deal of fresh information and the pattern of the Cosmos as a whole may be taking clearer shape. We may have discovered that space is infinite, and the galaxies extend onwards for ever—or we may have proved that it is curved and of limited volume, so that although the total number of galaxies will be immense it will nevertheless be finite. These questions of cosmology are, however, outside the scope of this book: let us return to our own galactic system, the Milky Way.

If we assume that only one sun in a thousand has planets—and this may well be a gross under-estimate—that would give a total of perhaps a hundred million solar systems in our Galaxy alone. Among all these, it can hardly be doubted that there would be many worlds on which life of some kind would be possible: there would even be a large number which would have physical conditions similar to those of Earth.

Since we do not know how life originated on our own planet, we can scarcely draw very useful deductions about its hypothetical existence on equally hypothetical other worlds. Yet if there is one thing that modern astronomy has taught us it is that there is nothing exclusive or peculiar about our place in the Universe. We may consider ourselves, and with some justice, to be the lords of the Earth—but we can hardly expect to find ourselves unrivalled in the whole of space, or even in our own corner of the Galaxy.

The geological evidence indicates that the Earth was formed about 3,000,000,000 years ago, and the other planets were probably born at the same time. Man has been on Earth for much less than a thousandth of this period, although as far as the climatic conditions are concerned human life could have flourished for the greater part of it. For some reason intelligence has appeared on the stage in the last second of a play that has already been running for an hour.

Elsewhere in the Universe—for that matter, possibly elsewhere in our own Solar System—it may have appeared much earlier. When one considers the history of mankind, and compares it with the ages of the stars, it is difficult to avoid the conclusion that ours must surely be one of the youngest races in the whole of space. It is of course impossible to be certain of this, but its *probability* seems very high. If one believes that life is a characteristic phenomenon in the Universe, and not a rare disease that has attacked a handful of unimportant worlds, then the conclusion is unavoidable that there must be at least *some* races far older and presumably far more advanced than our own.

The truth, or otherwise, of this proposition must be regarded as one of the central questions of philosophy. Yet it is one that has scarcely ever been faced, perhaps because its implications are too overwhelming—or too humiliating. Or is it because of the feeling that it can never be of any practical importance, because the stars must remain for ever beyond our reach?

This view is held even by some who are quite convinced of the possibility of *interplanetary* flight. It is, nevertheless, curious how often the subject of interstellar travel crops up among hardheaded engineers and scientists interested in astronautics—though it does so in an oddly indirect fashion. One seldom hears them say outright: "Oh yes, I believe that ultimately it may be possible to reach the stars as well as the planets." But if anyone attempts to prove the total impossibility of interstellar flight, there is a great show of indignation and calculations are promptly produced refuting the critics.

In the next chapter, we will consider just what is implied by travel to the stars. The reader may, if he wishes, regard this purely as an intellectual exercise—an exploration of ultimate possibilities, however unlikely they may be. But we believe that they are possibilities which *should* be explored, if only in an attempt to discover the final limits of astronautics.

17. To the Stars

There was a young lady named Bright
Whose speed was far greater than light
She set out one day
In a relative way
And returned on the previous night.

ANON.

TO send a spaceship to the stars, if the time factor is of no importance, requires little more energy than many interplanetary missions. Thus a rocket which left the Earth at a speed of fifteen miles a second would still have eleven miles a second of its original velocity left when it had escaped completely from the Solar System. If aimed in the correct direction, it would reach any of the nearer stars (ignoring, for the moment, the fact that the stars themselves are in motion at speeds of several miles a second). The journey to Proxima Centauri would last a trifle more than 70,000 years. Thus although a *very* far-sighted civilisation might consider it worth while sending messages to the nearer stars in the faint hope of getting some sort of reply a few hundred thousand years later, this is hardly what one has in mind when speaking of interstellar travel!

Clearly, speeds comparable to the velocity of light will be needed if even the nearest stars are to be reached in a human lifetime. As has already been stated, there are eight stellar systems within ten light-years of the Sun. Given unlimited power, there is no theoretical reason why such speeds should not be obtained: they are, however, as much beyond our reach today as present rocket speeds would have been beyond the attainment of the men who built the first steam-engines. The velocity of light is 670,000,000 m.p.h.—and we have seen how difficult it is to achieve the modest 25,000 m.p.h. or so needed for interplanetary travel. It is obvious that before interstellar travel can enter the realm of serious study, some new method of propulsion, coupled with a source of energy vastly more powerful than anything in sight today, will be required.

It is hardly likely that the rocket will always remain undisputed master of space: its disadvantages are so many and so obvious that if anything better comes along astronauts will seize upon it with considerable joy. Nevertheless it is just possible that some form of rocket propulsion might play a part in taking mankind, literally, to the stars. After the chemical and the atomic rockets will come the electric, or ion, rocket—a device whose outlines can be foreseen even today. It is possible, by appropriate electric and magnetic fields, to accelerate charged particles (ions) to velocities near that of light. A beam of such particles would act as a rocket jet and could produce a thrust—indeed, there is a well-known laboratory experiment to demonstrate this. With ordinary currents and powers the thrust is negligible, but there is no fundamental reason why such a device should not be developed for propulsion, particularly under the conditions prevailing in space.

To attain really high speeds by this means it would be necessary to have an extremely efficient source of nuclear energy: anything like our present atomic power plants would be completely useless. Some means would be required of releasing a large percentage of the energy of matter (not merely the 0.1 per cent. we can liberate today) and then transferring it, with negligible losses, to the ion beam. No way is known of doing this, but it does not involve any inherent impossibility. With the *total* annihilation of the "fuel"—i.e. its complete conversion into energy—we should have the extreme case of a rocket whose "exhaust" would consist entirely of light, the thrust of which under ordinary circumstances is far too small to be of any importance, though even an electric hand-torch gives an immeasurably small "kick" when switched on.

How fast the rocket itself would travel would depend on the

weight of fuel it carried, although owing to the Theory of
Relativity the laws now operating would be more complicated
than those explained in Chapter 3. When we allow for this, we
find that if we could build a rocket 90 per cent. of whose mass
consisted of fuel which was converted entirely into radiation,
the empty rocket would reach a speed equal to 98 per cent. of
that of light, or just over 600,000,000 m.p.h. At this speed, it
would reach the nearest star in four and three-quarter years.
However, we are dealing with such enormous velocities that
the period of acceleration (negligible in interplanetary flight)
must now be considered, and this would increase the time of
transit by a considerable factor. Indeed, on the "shorter" (!)
interstellar flights it would be necessary to start decelerating
again well before the speed of light had been approached.

As is now well known, at velocities approaching that of light
some very peculiar things begin to happen. These were pointed
out by Einstein when he formulated the Special Theory of
Relativity, and they have since been demonstrated experimentally
in certain cases. The effects which concern us here involve the
mass of an interstellar spaceship, and the "time-scale" of its
occupants.

The mass of a body increases with its speed. The effect is
immeasurably small at ordinary velocities, and is still negligible
at speeds of many thousand miles a second. But it becomes all-
important as one nears the velocity of light, and at that speed
the mass of any object would become infinite. This, of course,
is merely another way of saying that the velocity of light can be
approached as closely as one pleases but can never be actually
attained. (Like the absolute zero of temperature.)

In addition to the increase of mass, there is also what might be
called a "stretching" of time. If one could compare a minute,
as observed in the spaceship, with one on Earth, it would appear
to be longer. Again the effect is very small at ordinary speeds
but becomes enormous as one nears the velocity of light. At this
limiting speed, in fact, time would appear to stand still.

It will be seen that this relativity effect is working in the right direction, as far as the interstellar travellers are concerned. From their point of view, it reduces the duration of the voyage. However, it is important to realise that this time-contraction only becomes really significant at speeds greater than half that of light—i.e. over 300,000,000 m.p.h.

To take an actual example, consider the spaceship mentioned on page 176, which, by converting 90 per cent. of its mass into radiation, reached a speed equal to 98 per cent. that of light. Suppose it left Earth at an acceleration of two gravities and maintained this until it had reached its final speed. Then, from the viewpoint of the crew, the period of acceleration would have lasted just over a year. But from the viewpoint of an observer on Earth, the ship would have been accelerating for five and a half years!

Dr. Sänger, the noted German rocket expert, has made some calculations for an even more extreme case, purely as a matter of theoretical interest. He considers a spaceship *circumnavigating the Cosmos*—assuming that this represents a distance of 10,000,-000,000 light years. If the ship could achieve 99.999,999,999,-999,999,996 per cent. of the velocity of light, the crew would imagine that the journey lasted thirty-three years—yet 10,000,-000,000 years would have elapsed before they returned to Earth (if it still existed!). Since this feat would require the complete conversion into energy of a mass approaching that of the Moon, Sänger reasonably concluded that it surpasses the limits of what is ever likely to become technically feasible!

Without going to such an extreme case, however, it would seem that stars up to twenty or thirty light-years away could, in theory, be reached within a single human lifetime. The idea does not involve any mathematical or physical impossibilities, and the feat would not require too enormous amounts of fuel *if* the total conversion of mass into energy could be achieved. Although, at the moment, we have no idea how this might be done, it must be remembered that nuclear physics is still in its infancy.

Since several decades of travelling through space would be an ordeal even to the most enthusiastic astronauts, one cannot help wondering if medical science might come to the rescue with some form of suspended animation—another theoretical possibility still outside the range of present achievement. If this became practicable, it might extend the range of space-flight almost indefinitely—assuming that travellers could be found willing to return to Earth perhaps generations after their departure, when everyone they had ever known was dead and society itself might have changed out of all recognition.

Bernal, in the book already mentioned, has suggested that mobile artificial worlds, carrying whole populations on journeys which might last for centuries, may be used for interstellar travel. (In this sense, of course, we on this planet are interstellar voyagers already—though our journey is involuntary and we know neither its destination nor its beginning.) Such dirigible plane-toids might travel at a few per cent. of the velocity of light: there would be no appreciable time contraction at such speeds and after a trip round the nearer stars the fifth or tenth grand-children of those who began the journey would return to Earth. It is, however, difficult to see how a major fraction of the Cosmos could ever be explored by this means, since it would take some millions of years to travel from one end of the Galaxy to the other.

Even if one grants the existence of the necessary source of power, it may be asked if these enormous speeds are physically realisable owing to the existence of the interstellar gas mentioned on page 92, as well as of solid meteoric material in space. As far as the latter is concerned, calculations show that the danger is negligible, and it does not seem that interstellar hydrogen would be a menace except possibly at speeds not far short of that of light, when it might be necessary to consider some form of shielding.

The discovery of other forms of propulsion, not depending on the rocket principle, would not materially affect the above arguments. Perhaps one day, when we have learned something about gravitation and the structure of space, we may be able to produce one of the "space drives" so beloved of science-fiction writers. These drives, for the benefit of anyone unacquainted with contemporary mythology, have the great advantage that, since the force they produce acts uniformly on every atom inside the spaceship, accelerations of any magnitude can be produced with no strain on the passengers. Even if the ship were accelerating at a thousand gravities, the occupants would still feel weightless.

There is nothing inherently absurd about this idea: in fact, a gravitational field produces precisely this effect. If one were falling freely towards Jupiter, not far outside the planet's atmosphere, one would be accelerating at $2\frac{1}{2}$ gravities yet would be completely weightless. To take an even more extreme case, the very dense dwarf star Sirius B has a surface gravity at least 20,000 times as intense as the Earth's. Falling in such a field one would be accelerating more rapidly than a shell while it was being fired from a gun, but there would be no feeling of strain whatsoever.

If we can ever generate the equivalent of a controlled gravitational field we shall certainly have a very effective drive for spaceships—one which, combined with an appropriate source of energy, might enable speeds near that of light to be reached after relatively brief periods of acceleration. It would not, however, help us to circumvent the restrictions set by the Theory of Relativity. As far as any fact can be experimentally established, it seems certain that no material body can ever travel faster than light. This limit is a fundamental one, of a completely different character from the so-called "sonic barrier" which once seemed an obstacle to high-speed flight. There was never any doubt that one could fly faster than sound, *given sufficient energy*: the only problem was to obtain this energy, and like all purely technical problems this was eventually solved. But even the disintegration of all the matter in the Universe would not provide enough energy to enable a spaceship to reach the speed of light.

It appears, therefore, that when astronautical techniques have reached the limits set by the laws of Nature (and we must assume that eventually these limits *will* be reached, even though we are very far from them today) it will be possible to send expeditions to the nearer stars and for them to return within ten or twenty years—though the period of elapsed time might be considerably less to the travellers themselves.

To imagine any long-range exploration of the Universe we have to assume voyages lasting many centuries, or even millions of years. Such voyages could be possible only if whole generations were willing to exile themselves in space. This would not necessarily be a great hardship when one considers that a mobile planetoid would probably be a good deal larger—and would have incomparably greater facilities in every respect—than the State of Athens, in which small area, it may be remembered, a surprising number of men led remarkably fruitful lives.

It is, perhaps, worth pointing out that our present views on the subject of interstellar flight are conditioned by the span of human life. There is no reason whatsoever to suppose that this will always be less than a century. When we reach the stars, the achievement may be due as much to medicine as to physics. As Man's expectation of life increases, so a greater and greater volume of space will become accessible to exploration.

Before closing this chapter we must deal with two questions which any discussion of interstellar travel inevitably raises. In the first place, despite the categorical remarks made a few pages ago, can we be *absolutely* certain that the speed of light will never be surpassed? The Theory of Relativity is, after all, only a theory. May it not one day be modified, just as it modified Newton's law of gravitation, which had remained inviolate for centuries and was generally regarded as being absolutely correct?

Any attempt to answer this question would lead into the deep waters of philosophy, and would involve such ideas as the fundamental structure of space and time. It is doubtful if anyone alive today could contribute much of real value to such a discussion: the verdict must be left to the future. At the moment we can only say that the idea of speeds greater than light still belongs to that nebulous and faintly disreputable no-man's-land where may be found the Fourth Dimension, telepathy, and Dr. Rhine's disturbing experiments in para-physics.

The second question involves history rather than physics. If space-travel is possible, and if there are other intelligent races in the Universe, why have they never come to Earth?

One answer to this is given by Plate XIV(b). The tightly packed star-fields shown there cover only a small portion of the Milky Way, and each of those stars is separated from its neighbours by distances of several light-years. To travel from one side of the picture to the other, even at the speed of light, would take centuries—while the examination of every star and planetary system might occupy fleets of spaceships for millennia. It is therefore obvious that, even if a great many advanced civilisations were scattered around the Galaxy, we could not expect our own Solar System to be visited except at very rare intervals.

The situation would not be greatly affected even if the speed of light were not a limiting factor. A man may walk the length of a beach in a few minutes—but how long would it take him to examine every grain of sand upon it? For all we know, there may be fleets of survey ships diligently charting and recharting the Universe. Even making the most optimistic assumptions, it is hardly likely that our world (which lies, it will be remembered, in the thinly populated frontier regions of the Galaxy) would have been visited in the few thousand years of recorded history.

On the other hand, anyone who is willing to spend a lifetime browsing through old newspapers could collect an impressive amount of "evidence" for extra-terrestrial visitors, as was indeed done by the late Charles Fort. The record extends from the flaming dragons of ancient China to the "flying saucers" of today. Since it is, in the nature of things, never possible to prove that such apparitions did *not* come from outer space, the most reasonable attitude towards them would seem to be one of open-minded scepticism. It certainly seems very probable that there can be no other technically advanced civilisation in the Solar System *at the present moment,* otherwise we would indeed have had every reason to expect visitors. But during the millions of years that lie behind us—who knows? Life may have come and gone and come again on our neighbours while evolution here was working slowly towards Man.

Countless times in geological history strange ships may have drifted down through the skies of Earth, and left again with records of steaming seas, the first clumsy amphibians creeping upon the beaches and, much later still, the giant reptiles. A few of those ships may have come from other planets of the Sun, but most must have been strangers to our Solar System, travelling from star to star in their search for knowledge. And some day, they may return.

Even if we never reach the stars by our own efforts, in the millions of years that lie ahead it is almost certain that the stars will come to us. Isolationism is neither a practical policy on the national or the cosmic scale. And when the first contact with the outer universe is made, one would like to think that Mankind played an active and not merely a passive rôle—that we were the discoverers, not the discovered.

18. Concerning Means and Ends

Ah, who shall soothe these feverish children?
Who justify these restless explorations?
WALT WHITMAN—*Passage to India*

HAVING ranged, in imagination at least, throughout the Universe, let us now come back to Earth for a final summing-up of the position of astronautics today. Hitherto we have been concerned with purely scientific questions: now it is time to take the wider view.

That the conquest of space is possible must now be regarded as a matter beyond all serious doubt. It is a significant fact that, almost without exception, the outstanding authorities in the field of rocket propulsion have been enthusiastic protagonists of space-flight. This has been true of Professor Hermann Oberth, founder of European rocketry: of Dr. Eugen Sänger, who was in charge of the German Air Force's rocket research establishment at Trauen and is one of the world's leading experts on hypersonic flight: of Professor Wernher von Braun, Technical Director of Peenemunde, the birthplace of V.2.[1] In the face of such testimony, the pronouncements that space-flight is impossible, which are still occasionally made by experts in other fields of science, must be regarded as somewhat breathtaking examples of intellectual arrogance.

There is still, of course, room for a great range of opinions on details—and once again we would emphasise that many of the ideas put forward in this book must be regarded as *possible,* not *inevitable,* solutions to the problems of space-flight. If they are not adopted, it will be because something better has turned up in the meantime.

---

[1] It is on record that von Braun once remarked, in the early 1940s: "Oh yes, we shall get to the Moon—but of course I daren't tell Hitler yet."

Even its most enthusiastic supporters do not deny that the conquest of space is going to be a very difficult, dangerous and expensive task. The difficulties must not, however, be exaggerated, for the steadily rising tide of technical knowledge has a way of obliterating obstacles so effectively that what seemed impossible to one generation becomes elementary to the next. Once again the history of aeronautics provides a useful parallel. If the Wright brothers had ever sat down and considered just what would be needed to run a world air-transport system, they would have been appalled at the total requirements—despite the fact that these could not have included all the radio and radar aids which were undreamed of fifty years ago. Yet all these things— and the vast new industries and the armies of technicians that lie behind them—have now become so much a part of our lives that we scarcely ever realise their presence.

The enterprise and skill and resolution that have made our modern world will be sufficient to achieve all that has been described in this book, as well as much that still lies beyond the reach of any imagination today. Given a sufficiently powerful motive, there seems no limit to what the human race can do: history is full of examples, from the Pyramids to the Manhattan Project,[2] of achievements whose difficulty and magnitude were so great that very few people would have considered them possible.

The important factor is, of course, the motive. The Pyramids were built through the power of religion: the Manhattan Project under the pressure of war. What will be the motives which will drive men out into space, and send them to worlds most of which are so fiercely hostile to human life?

It is possible, as we have already seen, to list many excellent practical reasons why mankind ought to conquer space, and the release of atomic power has added a new urgency to some of these. Moreover, the physical resources of our planet are limited: sooner or later sheer necessity would have forced men to travel to

[2] The project which produced the material for the first atomic bomb.

the other planets. It may well be a very long time before it is easier—to take an obvious example—to obtain uranium from the Moon than from the Earth, but eventually that time is bound to come.

The suggestion has sometimes been made that the increasing pressure of population may also bring about the conquest of the planets. There might be something in this argument if the other planets could be colonised as they stand, but we have seen that the reverse is the case. For a long time to come, it is obvious that, if sheer *lebensraum* is what is needed, it would be much simpler and more profitable to exploit the undeveloped regions of this Earth. It would be far easier to make the Antarctic bloom like the rose than to establish large, self-supporting colonies on such worlds as Mars, Ganymede or Titan. Yet one day the waste places of our world will be brought to life, and when this happens astronautics will have played a major rôle in the achievement, through the orbital weather stations and, perhaps, direct climatic control by the use of "space-mirrors". When this has happened—indeed, long before—men will be looking hungrily at the planets, and their large-scale development will have begun.

Whether the population of the rest of the Solar System becomes ten million or ten thousand million is not, fundamentally, what is important. There are already far too many people on *this* planet, by whatever standards one judges the matter. It would be no cause for boasting if, after some centuries of prodigious technical achievement, we enabled ten times the present human population to exist on a dozen worlds.

Only little minds are impressed by size and number. The importance of planetary colonisation will lie in the variety and diversity of cultures which it will make possible—cultures as different in some respects as those of the Esquimos and the Pacific islanders. They will, of course, have one thing in common, for they will all be based on a very advanced technology. Yet though the interior of a colony on Pluto might be just like that of one on Mercury, the different external environments would inevitably shape the minds and outlooks of the inhabitants. It will be fascinating to see what effects this will have on human character, thought and artistic creativeness.

These things are the great imponderables of astronautics: in the long run they may be of far more importance than its purely material benefits, considerable though these will undoubtedly be. This has proved true in the past of many great scientific achievements. Copernican astronomy, Darwin's theory of evolution, Freudian psychology—the effect of these on human thought far outweighed their immediate practical results.

We may expect the same of astronautics. With the expansion of the world's mental horizons may come one of the greatest outbursts of creative activity ever known. The parallel with the Renaissance, with its great flowering of the arts and sciences, is very suggestive. "In human records", wrote the anthropologist J. D. Unwin, "there is no trace of any display of productive energy which has not been preceded by a display of expansive energy. Although the two kinds of energy must be carefully distinguished, in the past they have been . . . united in the sense that one has developed out of the other." Unwin continues with this quotation from Sir James Frazer: "Intellectual progress, which reveals itself in the growth of art and science . . . receives an immense impetus from conquest and empire." Interplanetary travel is now the only form of "conquest and empire" compatible with civilisation. Without it, the human mind, compelled to circle for ever in its planetary goldfish-bowl, must eventually stagnate.

It has often been said—and though it is becoming platitudinous it is none the less true—that only through space-flight can mankind find a permanent outlet for its aggressive and pioneering instincts. The desire to reach the planets is only an extension of the desire to see what is over the next hill, or

> Beyond that last blue mountain barred with snow
> Across that angry or that glittering sea.

Perhaps one day men will no longer be interested in the unknown, no longer tantalised by mystery. This is possible, but when Man loses his curiosity one feels he will have lost most of the other things that make him human. The long literary tradition of the space-travel story shows how deeply this idea is rooted in Man's nature: if there were not a single good "scientific" reason for going to the planets, he would still want to go there, just the same.

In fact, as we have seen, the advent of space-travel will produce an expansion of scientific knowledge perhaps unparalleled in history. Now there are a good many people who think that we have already learned more than enough about the Universe in which we live. There are others (including perhaps most scientists) who adopt the non-committal viewpoint that knowledge is neither good nor bad and that these adjectives are only applicable to its uses.

Yet knowledge surely is always desirable, and in that sense good: only insufficient knowledge—or ignorance—can be bad. And worst of all is to be ignorant of one's ignorance. We all know the narrow, limited type of mind which is interested in nothing beyond its town or village, and bases its judgments on these parochial standards. We are slowly—perhaps too slowly—evolving from that mentality towards a world outlook. Few things will do more to accelerate that evolution than the conquest of space. It is not easy to see how the more extreme forms of nationalism can long survive when men have seen the Earth in its true perspective as a single small globe against the stars.

There is, of course, the possibility that as soon as space is crossed all the great powers will join in a race to claim as much territory as their ships can reach. Some American writers have even suggested, more or less seriously, that for its own protection the United States must occupy the Moon to prevent it being used as a launching site for atomic rockets.

This argument (which reflects so faithfully the political paranoia of our times) fortunately does not bear serious examination.

The problem of supply—often difficult enough in *terrestrial* military affairs!—would be so enormous as to cancel any strategic advantages the Moon might have. If one wants to send an atomic bomb from A to B, both on the Earth's surface, then taking it to the Moon first would be an extremely inefficient procedure. Moreover, a lunar-launched missile could be detected a good deal more easily than one aimed from the other side of the Earth. A satellite in an orbit a few thousand miles high would seem to possess all the military advantages of the Moon, and none of its disadvantages. It would also be very difficult to locate, if it were covered with light and radar-absorbing paint.

It is one of the tragic ironies of our age that the rocket, which could have been the symbol of humanity's aspirations for the stars, has become one of the weapons threatening to destroy civilisation. This state of affairs has presented a difficult moral problem to those wishing to take an active part in the development of astronautics, for almost all research on rockets is now carried out by military establishments and is covered by various security classifications. The technical problems involved in designing long-range guided missiles are practically identical with those which will be met in the construction of the reconnaissance rockets described in Chapter 4. Separating the military and the peaceful uses of rockets is therefore an even more difficult task than creating atomic energy without atomic bombs.

This particular problem is not, of course, peculiar to rocket research: it can be encountered today in every field of scientific activity—even in medicine, for the power to heal is also the power to kill. It is, however, certainly more acute for the rocket engineer than anyone else except perhaps the nuclear physicist. He can only hope, if he thinks seriously about these matters (and scientific workers are no better and no worse than the rest of us in ignoring uncomfortable facts) that the results of his work will eventually be published and employed for peaceful ends.

This has already happened in the case of radar, which only ten years ago was top-secret yet is now used all over the world to provide safety at sea and in the air. It is true that the rocket has nothing like the immediate "civil" uses of radar. At the moment, indeed, it has only two non-military applications—high-altitude research and take-off assistance for aircraft. The ultimate and revolutionary uses of the rocket are all bound up with astronautics and are therefore still a considerable distance in the future.

There is little doubt that a great many scientists and engineers whose only interest in the rocket is as a means of crossing space have become involved in current military research because in no other way could they find the necessary support. It is worth quoting here some words written by Professor von Braun on becoming an Honorary Fellow of the British Interplanetary Society: "Is it not a shame that people with the same star-inspired ideals had to stand on two opposite sides of the fence? Let's hope that this was the last holocaust, and that henceforth rockets will be used for their ultimate destiny only—space flight!" Dr. Sänger expressed similar views on the occasion of his election: "If the great majority of human beings or the great organisers of human society were thoroughly convinced and enthusiastic about astronautics, then scientists and engineers could direct their research work immediately to space-flight problems. Unfortunately, it is not so. . . . Therefore, I consider the astronautical scientist's task is to turn human mentality slowly towards our target by steps of *fait accompli*. . . . Men are seldom convinced by good reasons, and more often by good facts."

Although, in present conditions, it may still seem a Utopian dream to hope for large-scale support for purely astronautical rocket research, with no military entanglements, it it not impossible that something like this may evolve in the future. When and if the political situation stabilises, and international co-operation on the scientific level is again resumed, the steadily growing astronautical societies in many parts of the world may be able, by their combined efforts, to act as catalysts and so bring about this desired state of affairs. This is one of the long-term plans behind the various International Congresses on Astronautics, the first of which took place in Paris during the autumn of 1950.

It should, however, be made quite clear that no society, as such, can now do effective large-scale research work in rocketry. The cost of a big rocket development programme is many millions of pounds a year: even a single model of a medium-sized liquid-fuel missile may cost several thousand pounds. The function of the astronautical societies, therefore, is not to attempt research and construction themselves—except perhaps into the various sub-sidiary problems which can be investigated without large budgets. Interplanetary societies will not build spaceships any more than aeronautical societies, *as such,* build aircraft. They will be the specialist organisations—the professional bodies—of the scientists and engineers doing work in this field.

When the time comes to build the first spaceships, the inter-planetary societies will be the spearhead of the attack: but their members will probably be acting under government orders—even though they may have had to persuade their governments to issue those orders in the first place!

It has sometimes been said that the main obstacles to interplan-etary flight are not technical, but political and economic. There is always an immense resistance to any change and a desire to pre-serve the *status quo.* Protagonists of space-flight frequently used to meet the remark: "Why go to the Moon? What's wrong with this Earth anyway?" Although the latter statement is seldom encountered these days, it has been succeeded by the query: "Why not devote all this effort to developing our own world before going to others?"

We have already given several answers to this question, point-ing out that many of the indirect consequences of space-travel will in fact help us to develop our own world—probably in in ways at least as unforeseeable as those in which the American oilfields and farmlands assisted the development of Europe. There is, however, a much more fundamental reply to this question, and one cannot help thinking that those who ask it have overlooked the facts of human nature. One wonders if they would have asked Pheidias, when he was starting work on the Parthenon frieze, why he was not engaged on something useful like rebuilding the Athenian slums. If he had kept his temper, the artist would probably have answered that he was doing the only job that interested him. So it is, in the ultimate analysis, with those who want to cross space.

There are, it seems, some people who have definite psychological objections to space-flight. In certain cases this has a religious basis—it is a new form of the old feeling that, in some mysterious way, there are things that "Man was never intended to do". We do not know a better way of demolishing this superstition than by referring to the old lady who remarked that aeroplanes were undoubtedly an invention of the Devil, "since men should travel in trains as God intended them to."

Others, one suspects, are afraid that the crossing of space, and above all contact with intelligent but non-human races, may destroy the foundations of their religious faith. They may be right, but in any event their attitude is one which does not bear logical examination—for a faith which cannot survive collision with the truth is not worth many regrets.

In the long run, the prospect of meeting other forms of intelligence is perhaps the most exciting of all the possibilities revealed by astronautics. Whether or not Man is alone in the Universe is one of the supreme questions of philosophy. It is difficult to imagine that anyone could fail to be interested in knowing the answer—and only through space-travel can we be sure of obtaining it.

We have seen that there is little likelihood of encountering intelligence elsewhere in the Solar System. That contact may have to wait for the day, perhaps ages hence, when we can reach the stars. But sooner or later it must come.

There have been many portrayals in literature of these fateful meetings. Most science-fiction writers, with characteristic lack of imagination, have used them as an excuse for stories of conflict and violence indistinguishable from those which stain the pages of our own history. Such an attitude shows a complete misunderstanding of the factors involved.

It has already been pointed out that ours must be one of the youngest cultures in the Universe. An analogy due to Sir James Jeans may help to emphasise this point. Take a penny, lay a postage stamp on it, and put both on top of Cleopatra's Needle.[3] The column then represents the age of the world, the coin the whole period of Man's existence, and the stamp the length of time during which he has been slightly civilised. The period during which life will be possible on Earth corresponds to a further column of stamps certainly hundreds of yards, and perhaps a mile, in height.

Thinking of this picture, we see how very improbable it is that the question of interplanetary warfare can ever arise. Any races we encounter will almost certainly be superhuman or sub-human—more likely the former. Only if we score a bull's-eye on that one stamp—indeed on a fractional thickness of that stamp —in the mile-high column will we meet a race at a level of technical development sufficiently near our own for warfare to be possible. If ships from Earth ever set out to conquer other worlds they may find themselves, at the end of their journeys, in the position of painted war-canoes drawing slowly into New York Harbour.

What, then, if we ever encounter races which are scientifically advanced yet malevolent—the stock villains, in fact, of that type of fiction neatly categorised as "space-opera"? In that event, astronautics might well open a Pandora's Box which could destroy humanity.

This prospect, though it cannot be ruled out, appears highly

[3] For the benefit of those unfamiliar with the Victoria Embankment or Central Park, this obelisk is about 70 feet high.

improbable. It seems unlikely that any culture can advance, for more than a few centuries at a time, on a technological front alone. Morals and ethics must not lag behind science, otherwise (as our own recent history has shown) the social system will breed poisons which will cause its certain destruction. With super-human knowledge there must go equally great compassion and tolerance. When we meet our peers among the stars, we need have nothing to fear save our own shortcomings.

Just how great these are is something we seldom stop to consider. Our impressions of reality are determined, far more than we imagine, by the senses through which we make contact with the external world. How utterly different our philosophies would have been had Nature economised with us, as she has done with other creatures, and given us eyes incapable of seeing the stars! Yet how pitiably limited are the eyes we do possess, turned as they are to but a single octave in the spectrum! The world in which we live is drenched with invisible radiations, from the radio waves which we have just discovered coming from Sun and stars, to the cosmic rays whose origin is still one of the prime mysteries of modern physics. These things we have dis-covered within the last generation, and we cannot guess what still lies beneath the threshold of the senses—though recent dis-coveries in paranormal psychology hint that the search may be only beginning.

The races of other worlds will have senses and philosophies very different from our own. To recall Plato's famous analogy, we are prisoners in a cave, gathering our impressions of the out-side world from shadows thrown upon the walls. We may never escape to reach that outer reality, but one day we may hope to reach other prisoners in adjoining caves, where we may learn far more than we could ever do by our own unaided efforts.

Yet space-travel will not, as some fear, destroy the mystery of the Universe. On the contrary, it may indeed increase it. Al-though many specific problems will be solved, and many doubts settled, our area of contact with the unknown will be enormously magnified. This has always been the case with scientific research: it should never be forgotten that, despite all our knowledge, we live in a far more wonderful and even more mysterious world than did our ancestors. We will not exhaust the marvels of the physical Universe until we have explored the whole Cosmos— and *that* prospect is still, to say the least, satisfyingly remote, if indeed it is theoretically possible. We have scarcely begun a voyage of discovery which may never have an end.

Somewhere on that journey we may at last learn what purpose, if any, life plays in the Universe of matter: certainly we can never learn it on this Earth alone. Among the stars lies the proper study of mankind: Pope's aphorism gave only part of the truth. For the proper study of mankind is not merely Man, but Intelligence.

Our survey is now finished. We have gone as far as is possible, at this moment of time, in trying to assess the impact of astronautics upon human affairs. Beyond this point the imagination can travel where it will, bounded only by the laws of logic.

I am not unmindful of the fact that fifty years from now, instead of preparing for the conquest of the planets, our grandchildren may be dispossessed savages clinging to the fertile oases in a radioactive wilderness. We must keep the problems of today in their true proportions: they are of vital—indeed of supreme—importance, since they can destroy our civilisation and slay the future before its birth. But if we survive them, they will pass into history and the time will come when they will be as little remembered as the causes of the Punic Wars. The crossing of space—even the sense of its imminent achievement in the years before it comes—may do much to turn men's minds outwards and away from their present tribal squabbles. In this sense the rocket, far from being one of the destroyers of civilisation, may provide the safety-valve that is needed to preserve it. Space-flight does not even have to be achieved for this to happen. As soon as there is a general belief in its possibility, that belief will begin to colour Man's psychological outlook.

We stand now at the turning point between two eras. Behind us is a past to which we can never return, even if we wish. Dividing us now from all the ages that have ever been is that moment when the heat of many suns burst from the night sky above the New Mexico desert—the same desert over which, a few years later, was to echo the thunder of the first rockets climbing towards space. The power that was released on that day can take us to the stars, or it can send us to join the great reptiles and Nature's other unsuccessful experiments.

The choice is ours. One would give much to know what verdict an historian of the year 3,000—as detached from us as we are from the Crusaders—would pass upon our age, as he looks back at us down the long perspective of Time. Let us hope that this will be his judgment:

"The twentieth century was, without question, the most momentous hundred years in the history of Mankind. It opened with the conquest of the air, and before it had run half its course had presented civilisation with its supreme challenge—the control of atomic energy. Yet even these events, each of which changed the world, were soon to be eclipsed. To us a thousand years later, the whole story of Mankind before the twentieth century seems like the prelude to some great drama, played on the narrow strip of stage before the curtain has risen and revealed the scenery. For countless generations of men, that tiny, crowded stage—the planet Earth—was the whole of creation, and they the only actors. Yet towards the close of that fabulous century, the curtain began slowly, inexorably to rise, and Man realised at last that the Earth was only one of many worlds; the Sun only one among many stars. The coming of the rocket brought to an end a million years of isolation. With the landing of the first spaceship on Mars and Venus, the childhood of our race was over and history as we know it began. . . ."

Index